State Making and Environmental Cooperation

Global Environmental Accord: Strategies for Sustainability and Institutional Innovation
Nazli Choucri, editor

Nazli Choucri, editor, *Global Accord: Environmental Challenges and International Responses*

Peter M. Haas, Robert O. Keohane, and Marc A. Levy, editors, *Institutions for the Earth: Sources of Effective International Environmental Protection*

Ronald B. Mitchell, *Intentional Oil Pollution at Sea: Environmental Policy and Treaty Compliance*

Robert O. Keohane and Marc A. Levy, editors, *Institutions for Environmental Aid: Pitfalls and Promise*

Oran R. Young, editor, *Global Governance: Drawing Insights from the Environmental Experience*

Jonathan A. Fox and L. David Brown, editors, *The Struggle for Accountability: The World Bank, NGOs, and Grassroots Movements*

David G. Victor, Kal Raustiala, and Eugene B. Skolnikoff, editors, *The Implementation and Effectiveness of International Environmental Commitments: Theory and Practice*

Mostafa K. Tolba, with Iwona Rummel-Bulska, *Global Environmental Diplomacy: Negotiating Environmental Agreements for the World, 1973–1992*

Karen T. Litfin, editor, *The Greening of Sovereignty in World Politics*

Edith Brown Weiss and Harold K. Jacobson, editors, *Engaging Countries: Strengthening Compliance with International Environmental Accords*

Oran R. Young, editor, *The Effectiveness of International Environmental Regimes: Causal Connections and Behavioral Mechanisms*

Ronie Garcia-Johnson, *Exporting Environmentalism: U.S. Multinational Chemical Corporations in Brazil and Mexico*

Lasse Ringius, *Radioactive Waste Disposal at Sea: Public Ideas, Transnational Policy Entrepreneurs, and Environmental Regimes*

Robert G. Darst, *Smokestack Diplomacy: Cooperation and Conflict in East-West Environmental Politics*

Urs Luterbacher and Detlef F. Sprinz, editors, *International Relations and Global Climate Change*

Edward L. Miles, Arild Underdal, Steinar Andresen, Jørgen Wettestad, Jon Birger Skjærseth, and Elaine M. Carlin, *Environmental Regime Effectiveness: Confronting Theory with Evidence*

Erika Weinthal, *State Making and Environmental Cooperation: Linking Domestic and International Politics in Central Asia*

State Making and Environmental Cooperation

Linking Domestic and International Politics in Central Asia

Erika Weinthal

The MIT Press
Cambridge, Massachusetts
London, England

Set in Sabon by Achorn Graphic Services, Inc. Printed and bound in the United States of America.

Library of Congress Cataloging-in-Publication Data

Weinthal, Erika.
State making and environmental cooperation : linking domestic and international politics in Central Asia / Erika Weinthal.
p. cm. — (Global environmental accord : strategies for sustainability and institutional innovation)
Includes index.
ISBN 0-262-23220-0 (alk. paper) — ISBN 0-262-73146-0 (pbk. : alk. paper)
1. Environmental policy—Asia, Central. 2. Environmental management—Asia, Central—International cooperation. 3. Environmental protection—Asia, Central—International cooperation. 4. Environmental degradation—Aral Sea Watershed (Uzbekistan and Kazakhstan). 5. Post-communism—Asia, Central. I. Title. II. Global environmental accords

GE190.A783 W45 2002
363.7'0526'0958—dc21

2001049200

Contents

Acknowledgments

Similar to the making of the Central Asian states, this book is embedded within a web of transnational actors and organizations. The initial idea of this project began with a summer grant from the Harriman Institute at Columbia University to study the political economy of cotton in Uzbekistan in 1992. Karen Peabody O'Brien braved the first Aeroflot flight with me to Toshkent. Throughout my journey along the rivers of Central Asia and at home in both the United States and Israel, the intellectual guidance and support of numerous people have shaped this project. In particular, I am greatly indebted to Jack Snyder, Barnett Rubin, and Steven Solnick for their knowledge and inspiration; more so, their sense of humor was greatly appreciated every time I sent them email from the field. Kate O'Neill and Pauline Jones Luong also read the manuscript in its entirety at its various stages. In Central Asia I was fortunate to have the friendship of Pauline Jones Luong; we shared many unforgettable moments and much laughter. Others that have taken the time to provide feedback and comments at the various stages of this work include Robert Bates, Valerie Bunce, Ken Conca, David Epstein, Rajan Menon, David Downie, Elinor Ostrom, and Brian Silver.

Many institutions provided funding and support for this project in Central Asia. In 1994 I participated in a Young Investigator Program on Water Resources Management with Turkmenistan and Uzbekistan sponsored by the National Academy of Sciences. The International Research and Exchange Board (IREX) funded my fieldwork in Kazakhstan, Kyrgyzstan, and Uzbekistan in 1995. Cassandra Cavanaugh was extremely helpful as the IREX representative in Toshkent. Besides Pauline and

Cassandra, Elizabeth Constantine and Derek Johnson were wonderful companions in Uzbekistan. The Ministry of Water in Uzbekistan and the Toshkent Institute of Engineers of Irrigation and Agricultural Mechanization served as my host institutions during my IREX grant. I am grateful to the many people in Central Asia who facilitated my research in the water, energy, and environmental sectors. They gave me much of their valuable time to explain to me the complex nature of water management in Central Asia. I am particularly indebted to the Avazmatov family, who opened up their home to me in Toshkent and on the state farm in the Fergana Valley. In Uzbekistan, the librarians at the Central Library in Toshkent and at the State Archives were extremely accommodating in locating materials. I am also appreciative for the assistance I received from the donor community and from the scholars who were always willing to share their expertise with me concerning Central Asian water politics—Jitzchak Alster, Michael Boyd, Arrigo di Carlo, Bill Davoren, Akhmal Karimov, Anatoly Krutov, Daene McKinney, Philip Micklin, Marcella Nanni, Werner Roeder, and Peter Whitford.

Other institutions that provided financial support to carry out and write up my results include the Graduate School of Arts and Sciences at Columbia University (through the Presidents Fellowship Program), the Carnegie Corporation Fellowship on Political Order and Conflict in the Soviet Union, the Foreign Language Area Studies Fellowship for Uzbek, and the Social Science Research Council (SSRC). Susan Bronson at SSRC deserves special thanks. The Center for International Security and Cooperation (formerly the Center for International Security and Arms Control) at Stanford University provided me with a warm and congenial home for 3 years. Aaron Belkin, Lynn Eden, Melanie Greenberg, David Holloway, and Donald Kennedy gave me much encouragement at Stanford University while I completed the final draft of my dissertation. The last place on this journey was the Department of Political Science at Tel Aviv University, where I was fortunate to have the first semester off to complete the manuscript revisions. Harry Leich at the European Division of the Library of Congress answered all my inquiries about how to best transliterate from the Central Asian vernacular into English. Finally, I thank Paul Bethge, Nazli Choucri, and Clay Morgan.

This project would not have materialized without the support of my family, who over the years had to learn where remote rivers and towns exist on the map. I am deeply grateful to my siblings, Benjamin Weinthal and Lois Weinthal. I thank my partner Avner Vengosh for his love and support. This book is dedicated to my mother, Silvia Weinthal, and in memory of my father, Henry Weinthal, who passed away shortly after I received my doctorate. He would have been pleased to see it in print.

State Making and Environmental Cooperation

1

The Aral Sea Crisis

You cannot fill the Aral with tears.
—Mukhammed Salikh, poet[1]

Control over water is power in Central Asia.
—Yusup Kamalov, Director, Union for the Defense of the Aral Sea and Amu Darya[2]

The Sea Is Dying

Central Asia is an arid environment in which three-fourths of the land mass is desert. The majority of the population lives in rural areas, concentrated in the oasis regions along the two main rivers: the Amu Darya (previously known as the Oxus) and the Syr Darya (previously the Jaxartes). These rivers originate in the eastern mountains of Central Asia and then flow across the Kara Kum and Kyzyl Kum deserts before emptying into the Aral Sea, a large terminal lake in the midst of the desert. For centuries, the territory between the two rivers was coveted by both the British Empire and the Russian Empire because the Great Silk Road ran through it. As a result of the struggle to gain access to Central Asia, British and Russian explorers generated numerous reports detailing the physical characteristics of the water basin and the economic activities of the local populations. In his account of reaching the shores of the Aral Sea with the Imperial Russian Geographical Society in 1874, Major Herbert Wood (1876, p. 186) observed: "Quantities of fish of large size sport in these foaming waters, over whose troubled surface flights of gulls and other aquatic birds hover and circle in search of their prey." In reference to the economic activity of the local population, Wood noted that "a great number of Karakalpaks are fishermen, who take, in fixed nets, quantities of a large, coarse sturgeon, with which the waters of the Amu abound, and which, dried and salted, form the staple of a very brisk trade carried on by the boats of the Amu and its branches, for distribution among the nomads of the Khwarezmian deserts and the sedentary populations of Central Asia" (ibid., p. 192).

More than 100 years later, the picture along the shores of the Aral Sea differed remarkably from Major Wood's description. While visiting the town of Muynak as the Soviet Union was collapsing, the Polish journalist Ryszard Kapuscinski (1994, pp. 261–262) captured the contrast:

> It is a sad settlement—Muynak. It once lay in the spot where the beautiful, life-giving Amu Darya flowed into the Aral Sea, an extraordinary sea in the heart of a great desert. Today, there is neither river nor sea. In the town the vegetation has withered; the dogs have died. Half the residents have left, and those who stayed have nowhere to go. They do not work, for they are fishermen, and there are no fish. . . If there is no strong wind, people sit on little benches, leaning against the shabby and crumbling walls of their decrepit houses. It is impossible to ascertain how they make a living. . . . They are Karakalpaks.

In only 30 years, the Karakalpaks have witnessed the drying up of the lake on which they had subsisted for decades. Although the Aral Sea was always saline, it supported a productive fishery. As the Soviet authorities withdrew water upstream for irrigation, the sea rapidly desiccated. With less water discharging into the Aral Sea, salinity increased from 10 grams per liter to more than 30 (Micklin 1992a).[3] Many of the native fish were unable to adapt to the rising salinity. As a result, commercial fishing came to a halt in the early 1980s. In 1959, the fishing boats and trawlers that now reside in the sand of the exposed seabed hauled in nearly 50,000 metric tons of fish (mostly carp, bream, pike-perch, roach, barbel, and a local species of sturgeon), but by 1994 the few fishermen that remained retrieved a mere 5000 metric tons of carp.[4] In order to keep the canneries operating and provide some form of economic sustenance for the affected local population, the authorities flew in fish from as far away as the Baltic Sea and the Pacific Ocean.

In short, under Soviet rule unprecedented amounts of water were diverted from the rivers to expand cotton monoculture and to reclaim new lands for agricultural production. These withdrawals for irrigation drastically altered the water balance in the Aral. The sea receded by 60–80 kilometers. Once the fourth largest lake in the world (behind the Caspian Sea, Lake Superior, and Lake Victoria), it shrunk to the sixth largest. In 1988 it bifurcated into a "small" sea in the north and a "large" sea in the south. Until 1960, about 55 cubic kilometers annually flowed into the sea. By the mid 1980s, the Amu Darya and the Syr Darya no longer emptied into the sea, which made commercial navigation practically impossible (Micklin 1992a; Micklin 1991).[5] (See table 1.1.) Between 1974 and 1986 the Amu Darya did not flow into the sea, and between 1982

Table 1.1
Year inflow of water from Amu Darya and Syr Darya (km³). Source: Rakhimov 1990, p. 9.

1960	56.0
1961	39.9
1962	35.1
1963	40.6
1964	51.7
1965	29.9
1966	42.8
1967	37.5
1968	36.3
1969	80.6
1970	38.5
1971	23.5
1972	22.6
1973	42.5
1974	8.2
1975	10.1
1976	10.3
1977	7.2
1978	19.7
1979	12.5
1980	8.3
1981	6.0
1982	0
1983	0
1984	4.0
1985	0
1986	0

and 1986 the Syr Darya did not reach the sea. In contrast, in the late 1800s the Russian colonialists relied on the Amu Darya (the Oxus) for navigation, which enabled them to fortify their strategic hold on Central Asia (then referred to as Turkestan). Yet by 1991 sea level had fallen by about 15 meters, surface area had been reduced by half, and the volume had diminished by two-thirds. In actual numbers, this meant that in 1960 the average area of the sea was 66,900 square kilometers; in 1991 it was 33,800. The average volume had diminished from 1090 cubic kilometers in 1960 to 290 in 1991 (Micklin 1992a, p. 275).

The water crisis became more pronounced in the 1980s, coinciding with indications of a severe economic and political crisis of the Soviet regime. Soviet authorities were no longer able to dismiss earlier warnings from the scientific community regarding the economic, environmental, and health consequences of the rampant and indiscriminate use of water for irrigation compounded by inadequate drainage. First, the cotton industry was in dire straits, as water logging and salinization of the soil were causing agricultural yields to decline even though production quotas from Moscow were increasing. Second, the quality of water in the rivers had deteriorated severely—especially in the Amu Darya, where until the 1960s the water was of satisfactory quality. Historically, the Amu Darya was the source of irrigation and drinking water for the populations of Khorazm Oblast'[6] and Karakalpakstan. By the mid 1980s, the small amount they received was laden with agricultural runoff containing large amounts of pesticides and herbicides, rendering it unfit for human consumption. Third, the desiccation of the sea led to a sharp upsurge in dust storms containing the toxic salt residue from the exposed seabed, and in place of the sea a new desert, referred to as the Akkumy (white sands), began to emerge (Smith 1994).[7] Finally, the downstream populations were unequivocally confronting a public health crisis as a result of the dust and salt storms and the contamination of the drinking water. Compounding the lack of potable water in the Aral delta, poor health conditions, inadequate diet, and high birth rates raised the rate of infant mortality to 75 per 1000 in Dashhowuz Oblast in Turkmenistan in 1988 and to 60 per 1000 in Karakalpakstan in 1989 (Micklin 1992b, p. 103).[8] In addition, there were numerous accounts of respiratory illness, esophageal cancers, typhoid, paratyphoid, and hepatitis among the populations bordering the Aral Sea (Carley 1989; Elpiner 1999).

The Russian geographer Arkady Levintanus (1992, p. 85) notes that "the desiccation of the Aral Sea is rightfully listed now amongst the worst ecological disasters of the twentieth century." For many, the desiccation of the Aral Sea ranks with the meltdown at Chernobyl as one of the worst examples of the Soviet Union's environmental legacy of utter wastefulness and unaccountability for human life (Ananyev 1989, p. 14). It is no wonder that, by the end of the Soviet period, the Soviet leadership was left with little choice but to officially designate the Aral Sea region a "zone of ecological catastrophe." The immediate cause of the water crisis was inefficient irrigation; however, the root causes of the Aral disaster were much deeper. Some suggest the underlying factors are related to the inappropriate strategy of economic development in Central Asia wherein Soviet planners emphasized agricultural raw products (primarily cotton, a water-intensive crop) rather than finished products or other traditional crops (Levintanus 1992; Rumer 1989). The Soviet economic system treated human beings and the environment as expendable for the sake of "progress." Preference was given to industrialization (and to heavy rather than light industry), mechanization, and economic specialization; as a result, the authorities blatantly ignored environmental protection and health and safety issues so that they could increase production in order to meet higher annual targets. The price was steep for diverting water to promote cotton monoculture. Indeed, the socio-economic choices made during the Soviet period succeeded in destroying a whole people's culture and livelihood, namely that of the Karakalpaks. In Karakalpakstan there is such a sense of hopelessness and fatigue among the population that when glasses are raised in honor of a foreign guest the locals regularly toast their "environmental poverty."[9] In like manner, Tulepbergen Kaipbergenov (a well-known writer from Nukus, Karakalpakstan) recalls:

> Now [Nukus] is a city filled with dust blowing about. But I remember very well how different it was. The air was different, the color was different, and life was different. Then practically, all the roads led to the Amu Darya, on which our city stood. . . . It was like that not very long ago. Thirty years ago, even less. And nothing from that remains today. A fishing village is in the past. The pier is in

the past. . . . The last time the Amu Darya or the *Zheihun* (the Furious River, as it was called in the past) inundated these places was in 1968. In 1971, the water already stood motionless. . . . An ecological catastrophe occurred and today continues along the Aral. . . . There are victims; there are people who for their whole lives are crippled.[10]

Internationalizing the Aral Sea Crisis

The breakup of the Soviet Union transformed a domestic water crisis into one of international relations for the newly independent Central Asian states. For the first time since the wave of decolonization in the 1960s, a major river system has undergone a process of political reorganization. The rivers that constitute the Aral basin became international rivers overnight. The Amu Darya extends across three new states (Tajikistan, Uzbekistan, and Turkmenistan), and the Syr Darya flows among four new states (Kyrgyzstan, Uzbekistan, Tajikistan, and Kazakhstan).[11] Although rivers physically unite their users, politically they demarcate borders. Accordingly, the introduction of new political borders had an immediate impact on the social, economic, and political relations of the 35 million persons living within the Aral basin. For downstream populations, such as the Karakalpaks, it became uncertain who now had the authority and the capacity to address the past ills caused by indiscriminate use of water for irrigation that had resulted in the "death" of the Aral Sea. Would Moscow follow through on the Soviet Union's commitment to help the Central Asian republics restore the Aral Sea, or would the newly independent Central Asian states have to figure out an appropriate solution to the water crisis alone?

The new Central Asian states are similar to other developing countries in that water demands are increasing rapidly as a result of high rates of population growth and an economy based on agricultural production. Without additional sources of water, the Central Asian successor states will not meet the basic needs of their populations in the twenty-first century. For economic and ecological reasons, cooperation is crucial for states that share an international river system. Sandra Postel of the Worldwatch Institute points out that cooperation is "essential not only to avert conflict but to protect the natural systems that underpin regional economies" (Postel 1996, p. 42). To prevent discord over water allocations and water quality, the Central Asian successor states must sustain cooperation while adapting politically to a new state system and physically to an international river system. Yet, with the breakup of the Soviet Union, cooperation over joint fresh-water resources in the Aral basin is no longer just a technical problem; it is now also a political one that ultimately links issues of environmental scarcity and degradation with the political, economic, and social challenges inherent in the transition from communist rule.

Conflict or Cooperation in the Aral Basin?

Owing to the imminent need to find a solution to the Aral Sea crisis, scholars and policy makers in and outside the region assumed that the unsettling of political and physical borders would intensify violent conflict and competition over land and water resources in Central Asia rather than engender the political conditions necessary for cooperation to take hold. The geographer David Smith (1995, p. 351) alleged that, since political borders no longer corresponded to the physical borders of the river system but now divided them, "nowhere in the world is the potential for conflict over the use of natural resources as strong as in Central Asia." Sergei Panarin of the Institute of Oriental Studies in Moscow conjectured that "the extreme shortage of water for irrigation is bound to bring to the fore, in an acute form, the issue of national control over water sources" (1994, p. 87). The World Bank concluded in a preliminary report (1993a, p. iv) that "in a region in which water is life and virtually nothing can grow without irrigation, the competition for water will be acute." Moreover, Ze'ev Wolfson, a specialist on Soviet environmental issues, purported that "with a tangle of economic and social problems against a backdrop of a depletion of such basic resources as water and fertile land, one must expect a further increase in political instability and conflicts throughout the entire area of Central Asia" (1990, p. 45).

The aforementioned predictions that conflict would ensue in the post-Soviet period were predicated on the upsurge in ethnic conflict in Central Asia that marked the last few years before the breakup of the Soviet

Union. For example, in June 1990 a violent conflict between two ethnic groups in Osh, the Kyrgyz[12] and the Uzbeks, claimed at least several hundred lives. During the previous year, Tajiks and Uzbeks quarreled over land and water rights in the Vakhsh Valley, deadly ethnic strife erupted between Uzbeks and Meskhetian Turks in the Fergana Valley, and Tajiks and Kyrgyz fought over land and water rights in the Isfara-Batken district along the border of their republics. In all these instances, the social unrest was due to shortages of land and water.

Yet acute conflicts over water resources did not arise after independence. In fact, the Central Asian successor states embarked on a path of cooperation. State breakup and the subsequent political demarcation of the water system created unforeseen possibilities for the Central Asian states (which for all intents and purposes resemble developing countries) to engage in coordinated efforts to mitigate threats from their ethnic and environmental legacies. The need for collective action to resolve the Aral Sea tragedy resonated with the Central Asian leadership. In a speech on the status of the Aral Sea, President Karimov of Uzbekistan said: "The problem is that our destiny was controlled by others. Now the time has come to take a serious approach to the task. . . . The fate [of the Aral] is inseparably linked with that of the independent states of Turkestan as a whole. . . . Therefore, Uzbekistan, Kazakhstan, Turkmenistan, Kyrgyzstan, and Tajikistan must create a single powerful international organization to solve the problems."[13]

On February 18, 1992, shortly after gaining independence, the Central Asian states signed the first of several interstate agreements regarding cooperation in the management, utilization, and protection of the interstate water resources of the Aral basin. In March 1993, in Qyzlorda, Kazakhstan, the heads of state signed an intergovernmental agreement on solving the problems of the basin. In January 1994 they approved an action plan for addressing the basin's dire situation and for broader social and economic development in the basin. In the autumn of 1996 they renewed their commitment to water sharing, signing the Nukus Declaration to strengthen the nascent international institutions for joint water management of the rivers. In March 1998 the Prime Ministers of Kazakhstan, Kyrgyzstan, Tajikistan, and Uzbekistan endorsed a limited water sharing agreement over the Syr Darya.

Environmental Cooperation among Transitional States

The purpose of this book is to explain why rapid regional environmental cooperation emerged where we would least expect to find it—between new states with a history of ethnic tension and over an international river system—and what form that cooperation took. The regional environmental cooperation that ensued in the Aral basin contrasts with the historical record, in which cooperative agreements over international river systems prevail more often in the developed industrialized countries than in developing countries (LeMarquand 1977). The unsettling of both political and physical borders and the creation of new states in the Aral basin raises the following questions: Broadly, how do new states embark on regional cooperation during periods of transformation? Why will new states agree to build interstate institutions before they have reconfigured domestic state institutions? How do new (and moreover transitional) states with weak institutional capacity deal with complex political and environmental problems? Under what conditions are these states able to negotiate institutional arrangements to overcome collective action problems in situations where the incentive structure precludes cooperation? Even if states succeed in cooperating over their shared water resources, will this form of cooperation be sufficient to improve the environmental situation?

Simply put, the Central Asian states must simultaneously engage in regional environmental cooperation at the international level and in state building at the domestic level. These are concurrent processes generated by the unsettling of physical and political borders. The puzzle presented by environmental cooperation among transitional states thus demands an integrative approach that connects domestic and international politics. In chapter 3, to explain interstate cooperation over the Aral basin, I develop an approach to two-level institution building that links environmental cooperation at the international level to state building at the domestic level.

Conventional approaches that are based on two-level games perceive states to be the main actors (Putnam 1988). My approach perceives international organizations (IOs), bilateral aid organizations, and nongovernmental organizations (NGOs) as the primary actors. At one table, these organizations must negotiate with the Central Asian governments to reach an international agreement on water sharing; at the other table, they must negotiate with the local communities hardest hit by the transitional period. Even though the overarching objective for these third-party actors or what can be considered transnational actors is to foster interstate cooperation in the Aral basin, they are entangled in the domestic game of state making in which side payments[14] are dispensed as inducements for regional cooperation that in turn are used by government elites to compensate key domestic constituencies that could undermine an agreement or threaten the government's hold on political and social stability. Thus, at the interstate level, side payments from third-party actors induce regional cooperation; at the domestic level, the introduction of side payments affects the structure of state formation.

With the end of the Cold War, the number of IOs and NGOs has increased tremendously. Similarly, the nature and the scope of their activity have broadened, making it necessary to investigate the precise role they play in world politics and the extent to which they have an impact on global issues and on the internal functions of states (Mathews 1997). The emerging literature on the internationalization of environmental protection (Keohane and Levy 1996; Schreurs and Economy 1997; Darst 2001) has helped to specify the growing influence of non-state actors such as IOs, NGOs, and multinational corporations in bringing about cooperation and collective action. Here, non-state actors define environmental issues, place them on the policy agenda, heighten awareness, mobilize domestic actors to push their governments to take action, and participate in monitoring and implementation (Kamieniecki 1993; Princen and Finger 1994; Porter and Brown 1991; Zürn 1998). Yet a smaller collection of researchers interested in the "pathologies" and/or the "perverse effects" of IOs and NGOs have also begun to focus on how third-party actors shape the internal functions of states or even relieve the state of its internal functions (Barnett and Finnemore 1999).[15] In Central Asia, IOs, bilateral development agencies, and NGOs assume this dual and sometimes contradictory role, in which they affect both interstate cooperation and state building through side payments.

This intervention in the internal affairs of new states creates a dilemma for IOs and NGOs. On the one hand, they help to maintain stability during this period of transformation and domestic flux in Central Asia; on the other hand, these sources of assistance allow corrupt members of the nomenklatura[16] to remain in place. In new states with weak domestic administrative structures, regional leaders can rely on previous institutional structures to secure domestic support and, as a result, can continue to appease local groups instead of building new national constituencies. As it turns out, in Central Asia the inchoate nature of domestic institutional structures permits national and regional elites to advance the short-term interests of their local constituencies in exchange for short-term payoffs of political and social stability.

Despite this paradox, without an overtly active role for IOs, bilateral aid organizations, and NGOs the Central Asian states may not have immediately established new institutions for regional cooperation; rather, other outcomes of discord or non-institutionalization may have transpired. IOs, bilateral aid organizations, and NGOs were able to replace the lost Soviet resource flows with alternative sources of financial and material assistance. Although these agreements may have mitigated violent ethnic conflict over scarce natural resources in the post-Soviet period, they certainly have not helped the Central Asian states to mitigate environmental degradation. The form of cooperation that has emerged in the Aral basin has reinforced social and political control rather than producing meaningful environmental protection.

Thus, in addition to focusing on how the active and purposive role for IOs, bilateral aid organizations, and NGOs has influenced whether or not the Central Asian states were able to cooperate, this book investigates the form that cooperation has and has not taken. Why were the institutions designed not the most environmentally efficient, even though they were the most politically efficient? In order to explicate why these new interstate institutions were unable to deal with the roots of the Aral Sea tragedy, in chapters 4 and 7 I explore the political and social remnants of the Soviet legacy of cotton monoculture, which continued to constrain Central Asian state building and regional environmental cooperation. Even when state breakup disrupted previous patterns of traditional rule based on patronage, the legacy of cotton monoculture enabled national and regional elites to maintain a strong hold on state power and social control in the Aral basin. By providing for a system for social control, cotton monoculture managed to impede any radical measures to effectively address the Aral Sea crisis; in fact, they directly influenced the process of state building. Similarly, the importance of the cotton sector as a source of foreign revenue has impeded attempts to reform agriculture and place it on the institutional agenda for those devising interstate institutions for the Aral basin. As a result, cooperation in Central Asia has been more about producing security regimes than about producing environmental-protection regimes.

The Nature of Transitional States

A central tenet of this work is that not all states possess the same capacity to deal with similar environmental problems. Developing countries, in particular, are worse at autonomously mitigating environmental problems in view of their lack of basic domestic capabilities. Conclusions generated by a research program on the linkages between environmental scarcity and acute conflict find that most scarcity-induced conflicts will be between states and will take place in the developing world (Homer-Dixon 1994, p. 19). Specific case studies and large-scale statistical studies have shown that it is often in developing countries where environmental factors are most likely to contribute to state failure and to increase the potential for internal conflict (owing to the weak ability of political and social institutions to absorb new stresses).[17]

Goldstone (1996, p. 70) has argued that in the field of environmental security what is needed is "better research on *what kinds of states* are likely to experience increased risks of failure due to population and environmental changes." Thus, in order to discern why cooperation may or may not emerge over an environmental issue and/or why the environment may or may not be a source of political instability and conflict, it is essential to adhere to such advice and to redirect the research agenda toward a focus on the kinds of states involved. By emphasizing the nature of states, this book contributes to mid-level theory building in the field of environmental politics, which in turn will help scholars and policy makers to predict better why environmental institutions may or may not meet the goals set by their designers.[18] With this in mind, we may then be able to design better strategies to counter environmental and physical changes in scarce resource systems and to prevent conflicts over resources. The case of the Aral basin presents scholars and policy makers with such a challenge, especially since finding a solution to the problem entails neither developed or developing states but rather post-communist states.

First, post-communist states are transitional states distinguished by their movement away from communism. In this context, the endpoint of the transition remains evasive—that is, it is not clear whether they will eventually become democracies. The Soviet system set out to integrate different societies and economies through centralization and hierarchy, but the post-Soviet period is defined by the dismemberment of their state socialist past. As part of the process of breaking ties to the past system of state socialism, these states must build political and economic institutions at the same time that they must reshape the national identity of the population. Indeed, when considering the economic and political transformations away from state socialism, we should not have expected the post-communist states to be better endowed to ameliorate resource scarcity and environmental degradation, insofar as they are poor and weakly institutionalized. Moreover, in the first few years after independence, the Central Asian states experienced, to varying degrees, periods of hyperinflation, rising unemployment, civil war, infrastructure collapse, pervasive corruption, deteriorating medical care, and declining living standards.

Second, with the end of the Cold War the post-communist states entered an international system dominated by a liberal economic order. The "triumph" of the free market and the absence of political, economic, and ideological alternatives to capitalist democracy gave the successor states of the Soviet Union and East Central Europe no choice but to embark on transitions toward a Western model.[19] As a consequence, the terms "democratization" and "marketization" cloak the transitions as these new states hope to join the North Atlantic Treaty Organization or the Organization for Security and Cooperation in Europe or (more important) to acquire coveted financial assistance from the International Monetary Fund and the World Bank. To meet the conditions set by Western IOs and bilateral aid organizations in order to receive aid, post-communist states are forced to hold elections even before domestic political parties and institutions are firmly established and to undertake economic austerity programs, which can result in greater income inequalities, higher unemployment, and rising local commodity prices. The Central Asian states have not been immune from these external pressures, which have also influenced the form and the scope of environmental cooperation in the region.

Most studies of cooperation and discord have focused on settled states; the theoretical literature has had few opportunities to consider states under conditions of transformation. In contrast to settled states, the challenges posed by the economic, political, and social transformations in the post-communist states for regional environmental cooperation are daunting in view of how weakly institutionalized and how poor they are. Yet it is the weakness of domestic institutions that is pivotal for understanding the likelihood of regional environmental cooperation in transitional states. In short, I argue that environmental cooperation in the Aral basin is nested within state making in Central Asia, which demands a theoretical explanation that links domestic politics with international relations.

Bridging the Gap between Domestic and International Politics

The interconnectedness between institution building at both the international and the domestic level in transitional states challenges the conventional literature on world politics that restricts international institutions as a subject for international relations and state making as a problem for comparative politics, even though it is frequently acknowledged that each of these processes transcends disciplinary boundaries. The separation is attributed to the different questions each discipline seeks to answer. Scholars of international relations are primarily concerned with the causes of foreign-policy outcomes and the nature of international politics, whereas comparativists concentrate on variations among domestic structures and state institutions. Rather than converge at the nexus of domestic politics and international relations, scholars have preferred to test domestic-level theories against those at the international level. Interaction effects between the two levels are seldom taken into account. As a consequence, causal arrows flow unidirectionally, resulting in second-image and second-image reverse analyses, for example. Second-image arguments focus on domestic sources of international cooperation that are derived from society-centered approaches, state-centered approaches, or approaches that link the state with society (Moravcsik 1993, p. 6). In contrast, second-image reverse arguments switch the causal arrows to explain domestic structure as a function of the international system such as a state's relative economic position in world markets (Gourevitch 1978).

Yet to understand cooperation problems for states under conditions of transformation we cannot restrict the analysis to either international causes or to domestic sources. Such reasoning from the international level to the domestic or from the domestic level to the international undermines the complex processes new states confront in periods of domestic transformation. Clearly, there is a need to fill this gap in the literature by connecting domestic processes with international ones.[20] One of the few attempts to merge domestic politics with international relations is through the development of two-level games in which domestic politics are an intricate part of international negotiations (Putnam 1988). Yet these approaches based on two-level games fail to include other actors that are not a constituent part of "the state" in the actual bargaining game.

Since the end of the Cold War, world politics is no longer just a game between states; it now entails multiple-level negotiations involving states, the international community, and domestic populations. Even in the case of the Aral Sea crisis, where the anthropogenic causes of the desiccation of the sea were well known, devising a solution required that international actors, national governments, and local populations participate in the process. By addressing two-level institution building, this book contributes to the broader theoretical literature on two-level games by highlighting the role that IOs, NGOs, and bilateral aid organizations assume in the negotiation process over new institutions for regional cooperation. By articulating the precise role that these third-party actors are playing at the level of regional cooperation and at the level of state building, my approach integrates domestic and international politics.

The Plan of the Book

Without being uncritically optimistic about the behavior of IOs and NGOs, I will analyze the mechanisms underlying their failures and their successes.

As this is a book about the politics of water, chapter 2 begins with the physical dimension of the Central Asian cooperation problem by depicting the various historical and topographical factors that influence governance over international river basins. In short, the physical makeup and the condition of a natural-resource system are the initial constraints on whether or not a resource becomes an issue of competition between users.

To establish the explicit cause-and-effect links of the general argument presented in chapter 3, I undertake an in-depth single-case study of the Aral basin. In chapters 4–6, I trace the process by which the international community influenced simultaneous institution building in the Aral basin at both the interstate and the domestic level through the use of side payments. In order to furnish empirical support for my argument, I draw on primary research I conducted in four Central Asian states: Kazakhstan, Kyrgyzstan, Turkmenistan, and Uzbekistan. During the period 1992–1998, I visited the region seven times and interviewed approximately 150 local and foreign water, energy, and agricultural experts. These interviews included meetings with official government representatives, international donors, local NGOs, and farmers. The broad scope of the interviews was necessary in order to evaluate what role each actor at the local, the national, or the interstate level was playing in building new institutions at both the domestic and the interstate level. By going back and forth between these levels, I was able to confirm or disconfirm the validity of the different actors' claims regarding the role of the international donor community and its impact on institution building in Central Asia. In addition to interviews, I relied heavily on on-site investigations in order to discern the local-level effects of decisions made at the interstate level and, in turn, how local institutions shaped interstate relations. Stays on several collective and state farms (kolkhozes and sovkhozes) in Uzbekistan (especially in the Fergana Valley), in Kazakhstan (the Shymkent region), and in Turkmenistan (the Dashhowuz region) and data garnered from Central Asian governments and from international organizations helped me to substantiate the importance of cotton monoculture as a form of social and political control. I combined these interviews and on-site investigations with library and archival research to document how water-sharing practices had changed in response to different external influences. Chapter 7 follows this presentation of the empirical data by looking at why certain institutions emerged and why others did not. In that chapter, I consider the different ways that the international community in conjunction with domestic actors could have constructed the Aral basin water game. I conclude with an examination of the unintended consequences of the role of the international community in Central Asia for environmental protection and for the early years of state building.

2

International Riparian Politics: Concepts and Constraints

Resource Scarcity, Environmental Degradation, and State Security

The tearing down of the Berlin Wall and the collapse of the Soviet Union added a large number of new states to the international system. The Soviet Union's collapse shifted the balance of power in the international system, culminating with the end of the Cold War. A narrow focus on military and ideological competition between the United States and the Soviet Union had dominated the Cold War era; the "environment" had been relegated to the realm of "low politics." In contrast, the profound changes in the structure of the international system sparked scholarly interest in other underlying causes and tensions that affect international security such as the relationship between resource scarcity and acute conflict.[1] Owing to these monumental events, many scholars argue that traditional understandings of international security should be reassessed and broadened to include human, physical, social, and economic well-being (Mathews 1989; Ullman 1983; Myers 1993; Homer-Dixon and Percival 1996; Kennedy et al. 1998).[2] Accordingly, environmental threats should be elevated to the realm of "high politics," as they affect not only the likelihood of conflict but also the well-being of individuals within states (Myers 1993).

Threats derived from water scarcity, in particular, could increase tension and generate conflicts between states. The heightening awareness concerning the effects of water scarcity and degradation on conflict and economic development clearly contributed to the widespread expectations that conflict would erupt over water resources in Central Asia immediately after independence. The logic is as follows: If water resources are scarce, competition for limited supplies then turns issues of both access to and quality of the water resource into a national security priority (Gleick 1993a). Homer-Dixon (1994), for example, finds that "the renewable resource most likely to stimulate interstate resource war is river water." Others, including Falkenmark (1986), suggest that water conflicts are more likely to occur in developing countries at the local and regional levels where water is critical for basic human needs and survival. Developing countries usually lack the domestic political capacity and financial resources to meet the challenges of stresses associated with overpopulation and poor water quality; hydrological conditions thus aggravate internal and interstate political tension.

Socio-economic factors such as population growth and economic development may exacerbate conflicts over water (Homer-Dixon, Boutwell, and Rathjens 1993).[3] In general, conflicts over environmental resources are much more complex than other traditional forms of conflict precisely because the linkages between resource scarcity and/or quality and conflict are muted by other social, economic, and political factors.[4] Solutions to environmental degradation and scarcity issues often involve tradeoffs with economic development; yet for developing and transitional countries environmental protection interferes with the expressed goal of promoting economic growth, as it carries high political and social costs. The states in the Aral basin are dealing with issues that affect economic development and environmental protection. If Uzbekistan, for example, were to decrease its dependence on cotton monoculture in order to release more water into the Aral Sea, that might lead to higher unemployment in the countryside and risk political and social instability. Indeed, the Central Asian states face a challenge similar to that faced by other developing countries: to reconcile environmental protection with economic growth.

The historical record, furthermore, underscores that interstate cooperation over shared river systems among developing countries is a rare phenomenon. In the volatile Middle East, interstate agreements on the Euphrates-Tigris, Yarmuk, and Jordan rivers are lacking.[5] The only formal allocative accord signed in the Middle East, the 1959 agreement between Egypt and Sudan over the Nile, excludes eight of the ten co-riparians (Waterbury 1994). At the same time, results from studies dealing with the direct and indirect links between water and conflict show that not all water disputes result in acute conflict.[6] Many stay at the level of heated rhetoric, and some even lead to protracted negotiations. Shared water resources can be both sources of conflict and sources of cooperation. Thus, as other fields in the social sciences have sought to explain the sporadic outbreaks of ethnic conflict by locating the conditions under which ethnicity becomes politicized, those who study water-resources issues need to ask similar questions concerning the conditions under which water turns into a source of conflict rather than of cooperation. Concerning international river systems, this entails a clear understanding of the physical parameters of the resource in question. An accurate picture of the physical constraints that shape the bargaining game over the formation of water-sharing institutions in Central Asia must also be embedded within the historical process of state breakup, since this rare political transformation has not only altered the nature of the state system but also affected the topographical and hydrological constraints within the Aral basin.

In short, this chapter sets the environmental context for bargaining over access to fresh water in Central Asia by highlighting the physical properties of international river basins generally and as they relate to the Aral basin specifically. In order to illustrate the physical constraints concerning whether or not the Central Asian states can achieve regional environmental cooperation, the first section enumerates why it is useful to frame international river basins such as the Aral basin as large-scale common-pool resources. In the second section, with this as a baseline, I focus on the specific constraints on bargaining over the formation of water-sharing institutions that entail a concatenation of topographical, hydrological, and historical factors. After laying out the concepts and constraints for building institutions in international river basins, I elucidate the various ways in which cooperation emerges over international river basins. Empirically, this chapter suggests that an enhanced role for third parties holds the key to fostering interstate environmental cooperation over the Aral basin. This point regarding a causal role for third-party actors is further elaborated in chapter 3.

Conceptualizing International River Basins

International rivers are by definition common to several states, and they usually cross and/or delineate state boundaries (Caponera 1992, p. 186).[7] According to a 1978 United Nations figure, there are more than 200 shared river basins, of which 13 are shared by five or more states and four are shared by nine or more states (United Nations 1978).[8] Shared watersheds account for 47 percent of the world's land area, and more than 60 percent of these are located in developing countries in Africa, Asia, and South America (Frederick 1996, p. 10). Many water conflicts have been and will be tied to this maldistribution of fresh water in the world. Most water-poor countries are also located in Africa, the Middle East, and parts of Asia. In many of these developing countries, rapidly growing populations reduce the availability of fresh water, especially as societies demand more water for agricultural, industrial, and household purposes.

According to Albert Garretson's foreword to Teclaff 1967, "river basins, despite their very great diversity in other respects, have one physical characteristic in common: each is a more or less self-contained unit within whose bounds all the surface and part or all of the ground waters form an interconnected, interdependent system." Similarly, Teclaff (1967, p. 3) identifies drainage—that is, moving water that flows down toward a single outlet—as the one characteristic common to all basins. The drainage basin in Central Asia extends across Turkmenistan, Kyrgyzstan, Tajikistan, and Uzbekistan, but covers only the two southern oblasts in Kazakhstan (Shymkent and Qyzlorda). The basin includes the water catchment areas of the following rivers: Amu Darya, Syr Darya, Zerafshon, Kashkadarya, Kafirnigan, Murghab, Tejen, Turgai, Sarysu, and Chu (ICAS 1996d, p. 7). In addition, a small part of the upper watershed is located in the mountains of Afghanistan and in Iran. On the whole, the rivers and their tributaries originate in the mountains of Tajikistan, Afghanistan, and Kyrgyzstan before flowing through the Kyzyl Kum and Kara Kum deserts to the Aral Sea. The Central Asian case primarily deals with surface water drainage rather than with a groundwater system, since the conditions in Central Asia favor surface water (owing to the presence of mountain glaciers that supply a large volume of meltwater). At the same time, this is a closed drainage basin, and thus any use of the runoff has an immediate impact on the salt balance of the inland sea.

After the introduction of new political borders in the Syr Darya basin, Kyrgyzstan was the upstream riparian, Uzbekistan and Tajikistan shared the middle course, and Kazakhstan (Shymkent and Qyzlorda provinces) was the downstream riparian. In the Amu Darya basin, Tajikistan became the upstream riparian; Uzbekistan was both a midstream and a downstream riparian in which the Autonomous Republic of Karakalpakstan and the province of Khorazm were the farthest downstream. The province of Dashhowuz in Turkmenistan was also a downstream riparian in the Amu Darya basin. The Zarafshon, a smaller drainage basin within the Aral basin, also had international consequences. The Zarafshon originates in Tajikistan and flows into Uzbekistan, where almost all of its waters are used in the provinces of Bukhoro and Samarqand before it vanishes in the desert, never reaching the Amu Darya. (Figure 2.1 is a map of the Aral Sea water system.)

When rivers are juxtaposed with states, two basic configurations are discernible: contiguous and successive international rivers (Caponera 1992, pp. 201–204). Each type of river creates different incentives for cooperation.[9] A contiguous river forms the boundary between two states, which means that the water cannot be used exclusively by one state at any time. In a successive river system, water flows through neighboring states and hence is used consecutively by them, which is the case for most international rivers. On the whole, water use depends on the location within the basin, and access to the water becomes exclusive within national boundaries. The geopolitics in successive systems, as is the Central Asian case, creates certain advantages for upstream states in such areas as flood control, water apportionment, and the ability to pollute. Downstream states, in contrast, may be restricted to granting navigation rights or to contributing to joint hydropower projects. Because of the unique advantage upstream riparian countries have over downstream riparian countries, the benefits of cooperation are highly asymmetrical and unevenly distributed. David LeMarquand, in a study of four international river systems (1977, p. 10), points out that there is "no economic incentive for cooperation when an upstream country uses an international river to the detriment of the downstream country and that country has

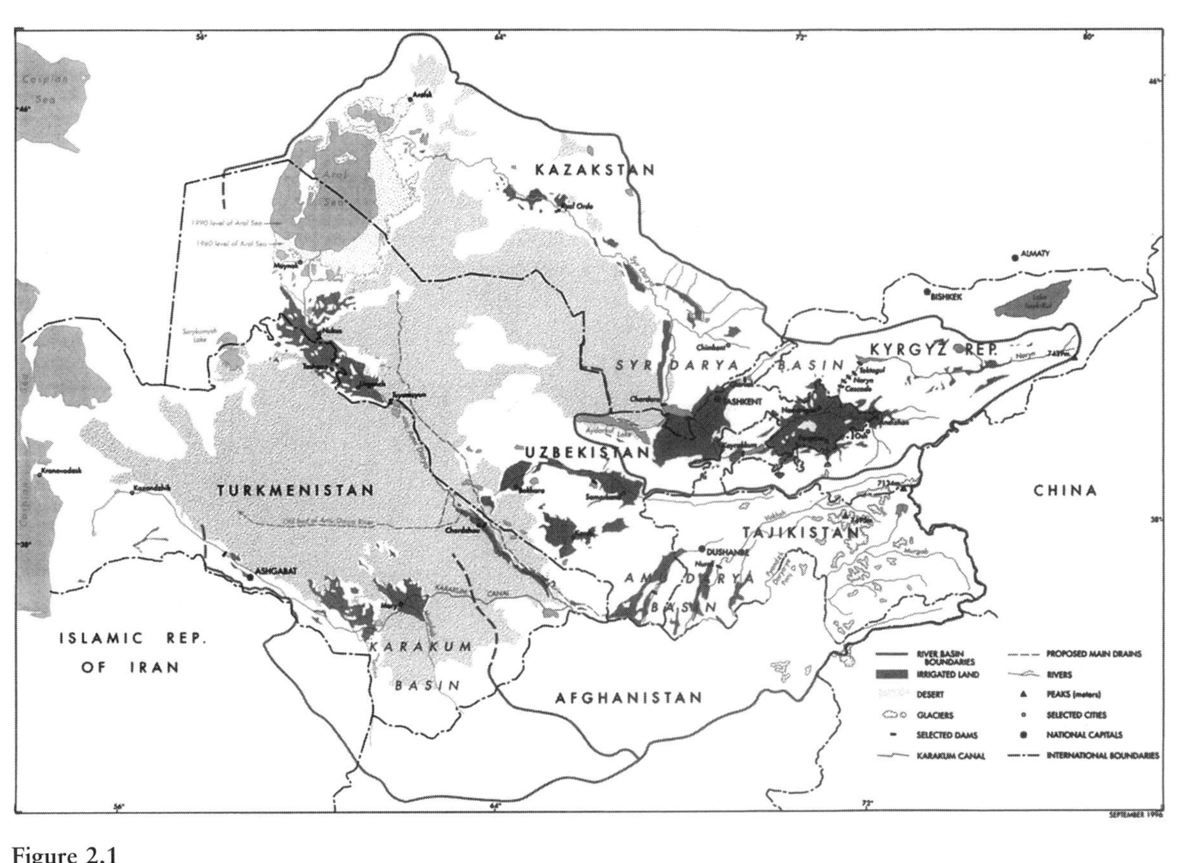

Figure 2.1
The Aral Sea water system. Source: International Bank for Reconstruction and Development.

no reciprocal power over the upstream country." Likewise, conflict is more likely when the downstream riparian is highly dependent on the river water and is the stronger state in the basin (Homer-Dixon 1994).

Thus, how an environmental resource is classified affects both the nature of the problem and the form of the solution. The Aral basin is a shared resource, an environmental system that extends across two or more distinct users.[10] Shared resources include nonrenewable resources (such as oil and gas) and renewable resources (e.g., water, fisheries, and forests). Controversies surrounding the utilization of shared resources involve the generation and distribution of externalities. Externalities are by-products of joint use. One widespread externality problem is the impact of upstream dumping of waste into a river in which the effects are felt far downstream from where the problem originated. Environmental externalities can increase discord among co-riparians even where overall relations are relatively stable and congenial. Even in Europe, where relations are friendly, discord arose between downstream and upstream riparians in the Rhine basin when the Netherlands found its share of the Rhine to be highly contaminated and polluted with chlorides dumped upstream (LeMarquand 1977, pp. 95–129).

In addition, the properties associated with shared resources and externalities exhibit many of the qualities of common-pool resources (CPRs). Ergo, it is useful to consider many of the conflicts associated with international river basins as CPR issues.[11] A common-pool resource is a "natural or man-made resource that is sufficiently large as to make it costly (but not impossible) to exclude potential beneficiaries from obtaining benefits from its use" (Ostrom 1990, p. 30).[12] If one focuses on the attributes of *excludability* and *subtractability* (Feeny et al. 1990), it is apparent that the Aral basin is not defined by open access. Indeed, it is costly for Kyrgyzstan to use water upstream exclusively for hydroelectricity in the winter, as this requires substantial amounts of financial and technical investments for infrastructure; at the same time, the downstream states cannot easily force Kyrgyzstan to refrain from exercising its sovereign right to use its water resources as it wills within its territorial borders. Yet, owing to Kyrgyzstan's withdrawals during the winter months, the downstream states feel the effects in the summer months, when their demand for irrigation water is high. Here, the benefits of the water basin are clearly subtractable (meaning that consumption by Kyrgyzstan in the winter months diminishes the amount of irrigation water available for the downstream users). Taylor (1987) characterizes this property of subtractability in terms of rivalry, since it is possible to deplete a resource through overuse and mismanagement (as happened with the fish that once thrived in the Aral Sea).

Research on common-pool resources is particularly useful for demonstrating how the scale of the resource influences the likelihood of cooperation (Young 1994b). Like other environmental problems, CPRs vary in scale; international rivers are clearly large-scale rather than small-scale CPRs. Since most small-scale CPRs do not transcend state boundaries or involve large heterogeneous groups of individuals, it appears easier to establish well-defined boundaries for them and to decide who has rights to use the resource (Ostrom 1990). Even though it is feasible, in principle, to demarcate the boundaries of an international river basin, and even to restrict access to it and to limit its flow through damming and water diversions, cooperation continues to be more unusual at the interstate level because no overarching authority or third party exists to enforce compliance. Rather, to sustain cooperation, a co-riparian must deliver on its promises and trust that other riparians will do likewise.

Regarding the question of externalities, most economic solutions mandate the internalization of externalities. Yet the issue of scale in an international river system precludes the internalization of social costs associated with economic activity, since political borders are incongruous with the physical borders of the resource. In fact, the creation of interstate water institutions for cooperation is unusually difficult on account of the large-scale nature of the resource in which multiple users with different capabilities and interests seek to appropriate a limited resource. Yet even when a river has been internalized (i.e., incorporated in a larger unit, as when the Aral basin was incorporated into the Soviet Union) this may not prevent overuse of the water system.

In summary: Cooperation over a large-scale CPR such as an international river system presents a problem of collective action for those trying to gain mutual benefits from the resource. It is not surprising that scholars have devoted considerable attention to the study of collective-action problems in international river systems, since the physical situation of an international river system precludes cooperation, or at least creates disincentives for cooperation. Among the many barriers to successful collective action in shared-resource systems are the shortage of institutions for joint management, free-rider and commitment problems, and the need for monitoring and enforcement mechanisms (Ostrom 1990).[13] Institution building for joint resource management is a highly contentious process, and the above discussion of the properties of CPRs and shared resources indicates at a minimum that empirically we should not have expected the process of rapid regional institutionalization in Central Asia that took place after the Soviet Union's breakup.

Constraints on Bargaining over the Formation of Water-Sharing Institutions

Simultaneous reconfigurations of political and physical borders such as occurred in Central Asia are rare. Redrawing borders in international river systems introduces new asymmetries of capabilities and interests between co-riparians, especially reinforcing the upstream-downstream divisions in a successive river system. By framing international rivers as common-pool resources, it is possible to grasp how the physical properties of water engender disincentives for cooperation. Although many of the impediments to water sharing were highlighted generally in the previous section in the context of excludability and subtractability of the resource, here I turn to the ways in which specific topographical and hydrological characteristics constrain the institution-building process for regional cooperation in an international river basin. In particular, the physical properties associated with large-scale CPRs generate conflicts of interests and of capabilities, or what some refer to as the "problem of heterogeneity" among the co-riparians.[14] As a result of the unsettling of political and physical borders that followed the collapse of the Soviet Union, the bargaining positions of the Central Asian successor states in the Aral basin are characterized by heterogeneous and shifting capabilities and interests.

History and Topography

John Waterbury's 1994 study of water cooperation and conflict in the Middle East identifies history and topography as the principal determinants of bargaining positions in an international river basin. Topography influences where people settle, and history allows certain patterns of use and appropriation to become customary. Combined, they determine how heterogeneous the interests and capabilities of the co-riparians will be.

The exploitation of a river basin is closely tied to the struggle of human beings to adapt to and transform the natural environment. Usually this process takes place first in the lower reaches of the basin, where the rich deposits of alluvial soils are suitable for settled agriculture. The practice of irrigated agriculture can be traced back at least as far as 4000 B.C. in Mesopotamia. With the intensification of water regulation and control, irrigation civilizations flourished in the Nile, Tigris-Euphrates, Indus, Yellow, and Yangtze basins (Teclaff 1967, p. 15). Around 1000 B.C., the advent of irrigation in Central Asia enabled both sedentary and nomadic populations to inhabit and cultivate artificial oases in the lower reaches of the Amu Darya and the Syr Darya (Khazanov 1992). Irrigated agriculture in the oases was a critical factor in shaping the ethnic and sociopolitical characteristics of Central Asia (ibid.).

Ordinarily, after settled agriculture is well established in the lower reaches, attempts are made to migrate into the upper reaches of the watershed. The intensification of late industrialization and urbanization creates incentives to exploit the upper watershed for hydroelectricity, for example. Hydroelectricity generation requires a much more sophisticated engineering technology than traditional irrigation systems that rely on flooding or channeling water through earthen furrows such as those found in large parts of the developing world. Only during the Soviet period were hydroelectric installations introduced into the upper reaches of the Aral basin in Kyrgyzstan and Tajikistan.

These competing needs and demands underpin the emergence of two different bargaining positions between upstream and downstream users over water appropriation. Although downstream users are usually the first to promote water applications, they are the weaker party when it comes to relative topographical position. In a situation where an upstream country wants to generate hydroelectricity, it may then try to refute a downstream user's claim to water rights by resorting to the principle of "first in use, first in right," which accords with the natural physical advantage that the upstream user possess. Downstream users, after all, will want to ensure that they receive the same quantity and quality of water as they historically have for irrigation and agricultural purposes, and will assert a position of acquired rights.[15]

As an illustration, it is hard to imagine that Ethiopia, an upstream state in the Nile basin, will continue to adhere to the status quo of established practice set forth by the 1959 treaty between Egypt and Sudan. While Ethiopia was mired in a civil war, it did not seek to develop the Blue Nile for its own agricultural development; with the cessation of hostilities, however, Ethiopia might decide to challenge the agreement between Egypt and Sudan. Any subsequent actions taken by Ethiopia upstream could dramatically reduce the supply of available water to the downstream countries. At the same time, it is even more doubtful that Egypt will wittingly relinquish what it considers to be its historical rights to use of the Nile, and as a more economically and politically powerful country it may be able to prevent Ethiopia from taking steps to develop the Blue Nile. Turkey, in contrast, has managed to change the status quo by unilaterally acting to restrict the flow of the Euphrates through the construction of large hydroelectric installations along the Euphrates; it has for all intents and purposes exercised its absolute topographical advantage. Turkey's interests and capabilities remain clearly inimical to Syria's and Iraq's position on how best to divide the waters of the Euphrates.

In view of this upstream-downstream tension, what does the introduction of new political borders mean for a previously domestic and highly integrated water system? The unanticipated breakup of the Soviet Union and reconstitution of East Central Europe complicates matters, since new political borders transform historical patterns of appropriation and utilization by altering the relative capabilities and interests among states in a river basin. New political borders, especially where none had previously existed, can empower nascent upstream states for the first time to exercise their claims to sovereign rights and to exclude or to limit downstream states' access to the river system. Indeed, Kyrgyzstan, after achieving independence, sought to exploit its hydroelectric potential by developing additional reservoirs along the Naryn River, a tributary of the Syr Darya.

Likewise, after Slovakia seceded from Czechoslovakia the Slovak government decided to press ahead with the completion of the Gabcikovo part of the Gabcikovo-Nagymaros project in the Danube basin as an overt symbol of Slovak sovereignty.[16] In short, state breakup can threaten historical patterns of water sharing, since the introduction of new political borders shifts the availability and quality of water supplies among users.

Hydrology

Roughly 97 percent of the water on the earth is salt water and thus is not readily available for drinking or agricultural purposes. Only 2.5 percent of the remaining water stocks are fresh water, but even these are unevenly distributed spatially and temporally (Shiklomanov 1993, pp. 13–24). Two-thirds of these fresh-water resources are locked in glaciers and ice caps. This limited supply of fresh water must meet the demands of multiple users in different sectors of society for both consumptive and non-consumptive purposes. Non-consumptive uses include navigation, hydropower, recreation, and fishing; consumptive uses include drinking and household, industrial, and agricultural activities. Water-consumption patterns depend on the particular climate conditions, lifestyle, culture, technology, diet, and wealth of the society in question (Gleick 1996, p. 83). As such, it is primarily the consumptive uses that place pressures on demand and create scarcities among competing users for limited, albeit renewable, water resources.

Aside from the relative topographical position of a country, an array of hydrological conditions associated with the availability of fresh water generate conflicts of interests and capabilities among co-riparians. Factors include the variability and uncertainty of water supplies, the degree to which the water supply is shared, and the perception of scarcity (Frederick 1996).

Gleick (1993a, pp. 99–104) identifies four quantitative indices for measuring the vulnerability of states to water-related conflicts: ratio of water demand to supply by country, per-capita availability, dependence on imported surface water, and hydroelectric dependence.

Per-capita water availability[17] is a good place to start. Many agree that a moderately developed country's absolute per-capita water availability should not fall below 1000 cubic meters per year. As of 1990, eighteen countries fell below this minimum requirement and could be considered water stressed (Gleick 1993a). Most of these countries are in Africa and in western Asia. Several of these water-stressed countries (Algeria, Burundi, Kenya, Rwanda) have experienced state collapse, civil war, and/or domestic turmoil. According to projections, more and more people will live in countries where water is scarce—by one estimate, between 1 billion and 2.4 billion (13–20 percent of the world's projected population) by 2050 (World Resources Institute et al. 1996, p. 302).

How water is used is another useful indicator of a state's interests and preferences in a bargaining situation over water-sharing institutions. Worldwide, agriculture consumes by far the largest amount of annual runoff: two-thirds (Shiklomanov 1993; Postel 1993). Irrigated agriculture is a substantial part of economic activity in many developing countries. Without irrigation to reclaim the desert sands, Central Asia would have been unable to produce 90 percent of the Soviet Union's cotton, a third of its fruit, a fourth of its vegetables, and 40 percent of its rice (Micklin 1991, pp. 10–11). Turkmenistan and Uzbekistan were the primary recipients of most of the large-scale irrigation projects that turned the deserts into fields for the production of cotton.

Owing to the physical interdependence of a river system, a country may find its preferences constrained and its internal decision making obstructed if a large fraction of its water supply originates outside its borders. The extent to which a resource is shared has an immense effect on the bargaining position of a riparian country. The fact that 97 percent of Egypt's water originates outside Egyptian territory clearly makes that country vulnerable to upstream actions by Sudan or Ethiopia. Hungary is extremely vulnerable to reductions in the flow of the Danube, since 95 percent of its water supply arises outside its borders (Gleick 1993a, p. 103); hence, it should not have been unexpected that Hungary would vehemently protest Slovakia's decision to divert a proportion of the flow of the Danube to complete the Gabcikovo dam.

State breakup and decolonization are rare, but when they happen they can wreak havoc on the governance of the physical environment. The unsettling of political and physical borders can change who controls the water. In regions where water is scarce, states located at the headwaters then gain a substantial amount of bargaining leverage. In Central Asia, the effects of the shifting of boundaries are immediately evident. Since independence, 98 percent of Turkmenistan's and 91 percent of Uzbekistan's water supply originates outside their borders (Smith 1995, p. 361). Even within Uzbekistan, some provinces (e.g., Andijon in the Fergana Valley) are left without any indigenous water supplies. All of Andijon's water comes not only from outside the region but from outside the republic (ibid.). Such basic indicators suggest that Central Asia could be susceptible to interstate conflict, insofar as the new states that are really water poor—Uzbekistan and Turkmenistan—are also the ones most dependent on the water resources from outside their borders for irrigated crop production.[18] (See table 2.1.)

Table 2.1
Dependence on surface water (percent of total flow) originating outside national borders of selected countries. Sources of data: Gleick 1993a; Smith 1995.

Turkmenistan	98
Egypt	97
Hungary	95
Mauritania	95
Botswana	94
Bulgaria	91
Uzbekistan	91
Netherlands	89
Gambia	86
Cambodia	82
Syria	79
Sudan	77
Niger	68
Iraq	66
Bangladesh	42
Thailand	39
Pakistan	36
Jordan	36
Israel[a]	21

a. The number for Israel includes only flow originating outside current borders, which could change with a political settlement to the Israeli-Palestinian conflict.

A country's dependence on hydropower affects its relations with its neighbors; it also affects perceptions of its internal sovereignty. Water may be the only valuable resource of a mountainous upstream country such as Nepal or Kyrgyzstan. Unlike the downstream states in Central Asia, which are poor in water but rich in energy, Kyrgyzstan lacks indigenous supplies of oil and gas. When Kyrgyzstan runs its hydroelectric plants in the winter, it reduces the water flow to Uzbekistan in spring and summer, when the demand for irrigation is at its peak downstream. Upstream states that are rich only in water perceive the ability to build and operate hydroelectric plants as a means of exercising sovereignty in internal economic policy making. Yet, because the resource is subtractable, the downstream states receive less water, which is also of a poorer quality than what is used upstream.

Hydrological conditions, like political factors, are dynamic. Scarcity and externalities in CPRs result directly from human interaction with the water resource through appropriation and utilization. Unsettling and reconstitution of political borders can shift hydrological relations among states. Therefore, the highlighting of several conventional indicators for water vulnerability illuminates how state breakup affects water availability and, in turn, shapes a state's interests and capabilities in bargaining over new institutions for water sharing.

The conjuncture of these hydrological, topographical, and historical factors helps us to comprehend why many expected conflict in Central Asia: The removal of the hegemon in Moscow might have created conditions ripe for conflicts over the competing uses for water since there are clear differences in downstream and upstream interests and capabilities. It also helps us understand the dynamics of bargaining over the formation of new water-sharing institutions.

Patterns of Cooperation over International River Systems: The Need for Third Parties

Building institutions for water management embodies the fundamental question of politics, as Robert Chambers (1980, p. 33) observes in reference to irrigation systems: "A central and universal issue in the distribution of irrigation water is who gets what, when, and where." Individuals and governments specifically design water institutions to provide the rules of the game for utilization and appropriation of a shared water resource in order to prevent its overuse and degradation. Water institutions mediate among human beings, the natural environment, and technology. Yet, in comparison with public-good arguments in which institutions are solutions to market failures, institutional solutions to CPR problems are much more complex, owing to the perception of scarcity and the asymmetry of interests and capabilities among the multiple actors involved.

The newly independent Central Asian states needed to cooperate over a large-scale water basin. The breakup of the Soviet Union exacerbated a situation of heterogeneity of interests and capabilities by removing the hegemon that had made a coerced equilibrium stable for so long. It disrupted the mutual interdependence or natural unity of the Aral basin with the introduction of new political borders. In order to effectively share the rivers of the Aral basin, the Central Asian states sought to re-create a situation of mutual interdependence whereby no one state will have any incentives to unilaterally use the water system (exclude others) that could cause harm to the other riparians (increase the level of rivalry). As a result, they needed to customize the provisions of a new cooperative water regime to correspond with the political reality of independence, since a centralized water authority in Moscow was no longer a viable or desired option.

This section examines conventional theories of cooperation in international relations to see what they would predict about interstate cooperation considering the physical dynamics of a water system. Integrating the political and the physical components explicates why regional water cooperation emerged rapidly in the Aral basin. What are possible patterns of cooperation in view of the fact that the benefits of cooperation are highly asymmetrical and unevenly distributed in an international river basin? How does the particular form of heterogeneity or homogeneity within a shared-resource system affect the likelihood of interstate cooperation? Three patterns of cooperation are possible in shared-resource systems: coerced, voluntary, and induced.[19] The combination of new states and heterogeneity within the resource system, moreover, suggests that induced cooperation, with an enlarged role for third parties, is pivotal for resolving the problem of water sharing in the Aral basin.

Coerced or Centralized Cooperation

Realists argue that cooperation or agreement emerges because a hegemon has sufficient structural power to either compel or coerce other states.[20] Coerced cooperation occurs where there are stark asymmetries of power or what others have denoted as an extreme form of heterogeneity (Martin 1994). In the domestic policy realm, such forms of coercion translate into recommendations for centralized decision-making, regulations, and/or private property.[21]

Because Moscow was able to mitigate conflicts among the Central Asian republics over water, and hence act a mediator, Moscow solved the problem of collective action among the Central Asian states. The Soviet government provided domestic stability and ensured cooperation in water sharing through the creation of an integrated plan for water management. At the same time, Moscow did not act as a benevolent hegemon, since it did not seek to preserve the water's quality. The Soviet case, in fact, demonstrates hegemons can be environmentally exploitative while providing a collective good in the form of political and social stability.

Solely in regard to the physical situation of an international river system, realist arguments posit that upstream states can force solutions for water sharing on downstream states, or that they can simply exploit a water system as they desire. When the asymmetries of capabilities and interests among riparian countries line up so that the most economically and militarily powerful country also has the most physical control of the resource (being upstream of the others), the upstream country can be expected to impose its will on the others regarding water allocation.[22] This was in fact what took place in January 1990 when Turkey unilaterally interrupted the flow of the Euphrates for an entire month to fill the huge reservoir behind the Atatürk Dam. Turkey's decision to build 21 dams along the Euphrates as part of the overall Southeast Anatolia Project to increase its hydroelectric potential and the amount of land available for agriculture heightened political tension with Syria and Iraq. The completed project will greatly restrict the flow of the Euphrates for Syria and Iraq, which are highly dependent on that river. With its profound topographical advantage, Turkey has been able to unilaterally determine the flow of the river. At the same time, Turkey's actions have precluded a comprehensive agreement or institutional arrangement among the co-riparians.

But do upstream states really control the water flow? Have there been instances in which an upstream state has been unable to impose its will on a downstream one? Most water systems are not governed entirely by such stark asymmetry or heterogeneity. Evidently, topography may not be the sole determinant of bargaining power; although an upstream position is a form of structural power, it does not automatically translate into bargaining leverage.[23] Other political, economic, and cultural factors can block an upstream state from taking such unilateral action or from purely dictating the terms of water sharing.

Even where an upstream riparian is more powerful economically and politically, as the United States is in relation to Mexico in the Colorado basin, other factors influence its bargaining position. LeMarquand (1977, pp. 25–47) argues that the United States agreed in August 1973 to decrease the level of salinity in the Colorado River for reasons tied to image, law, and linkage. Since it is difficult to always translate structural power into outcomes, hegemony as a solution to collective-action problems is an unusual case in world politics; moreover, stark asymmetry itself is rare. In Central Asia after independence, a hegemon no longer existed to dictate the terms of a water-sharing agreement; nonetheless, new asymmetries were created between upstream and downstream states with the introduction of political borders.

Voluntary or Decentralized Cooperation

Institutionalists are much more optimistic than realists about the ability of individuals and governments to avoid the "tragedy of the commons" in a CPR (Young 1994a; Taylor 1987).[24] Institutionalists consider overreliance on a centralized authority or a hegemon an impediment to efforts by individuals (at the local level) or by states to change the structure and the incentives of the situation in which they find themselves. Keohane and Ostrom (1994) argue that institutions emerge from voluntary contracting rather than from coercion or a centralized agency. Haas, Keohane, and Levy (1993, p. 4) find that world government is not an option; rather, they suggest, "organized responses to shared environmental problems will occur through cooperation among states, not through the imposition of government over them."

Some observers assert that physical interdependence leads actors to demand institutions for the joint management of a CPR. According to

Keohane (1982) and Oye (1986), institutions fulfill specific functions; they generate information, lower transaction costs, increase transparency, and reduce uncertainty. Ergo, institutions are by definition efficient. On this view, individuals and governments have well-defined preferences, enter negotiations knowing what they want, and are able to anticipate the effects of institutions. Concerning CPRs, the assumption is that, when individuals are able to devise their own institutions without any external help, they, as the appropriators of a resource, best understand what incentive structures should be incorporated into the institutional design in order to prevent conflict and overuse of the resource. However, in unsettled situations, as in Central Asia, preferences and interests are also being constituted throughout the negotiations, owing to concerns about state building.

Numerous case studies on local CPRs indicate that the most successful cases of cooperation are situated in small, homogenous communities. Homogeneity, social capital, local know-how, trust, and community encourage decentralized forms of cooperation over shared resources.[25] For example, attempts are made to involve a community in decisions concerning the design, the construction, the operation, and the maintenance of an irrigation system (Ostrom 1992). "Irrigation associations" or "water user associations" are propounded as the principal solution to water mismanagement problems.

Whereas heterogeneity at the local level creates obstacles to collective action (Lipecap 1994), realists view stark asymmetry as an asset for interstate cooperation (Martin 1994). Such discrepancies concerning the form of heterogeneity between the local level and the interstate level brings us back to the issue of scale; they also raise the problem of transporting theories of cooperation over small-scale CPRs to the international level and vice versa. Young (1994b) points out that it is not as easy to apply the term "community" to the international system as to apply it to small-scale societies. Accordingly, many of these voluntary and decentralized approaches to institution building appear to fit better with smaller irrigation projects. In Central Asia, Soviet planners completely re-engineered a decentralized water system to meet the demands of centralized planning.[26]

In summary: The CPR literature and institutionalist theories emphasize voluntary contracting when explaining institution building for cooperation over an international river basin. Under specific circumstances whereby co-riparians have symmetrical capabilities and/or interests, states may then be able to maintain the mutual interdependence of the resource system. Yet the basic physical structure of a successive river entails asymmetry between upstream and downstream riparians, which makes voluntary cooperation unlikely. Instead, riparian countries often need to figure out how to equalize the asymmetries in the benefits of cooperation. This may entail the use of voluntary compensation. In the Rhine basin, the Netherlands, France, Germany, and Switzerland agreed to share (although unevenly) the costs of the cleanup. Here, voluntary cooperation has been largely successful because interactions take place in a heavily institutionalized setting in which the countries are economically interdependent and relations among them are fairly good (Bernauer 1997). Moreover, these European states often have similar interests even though their physical capabilities might differ as a result of the upstream-downstream dichotomy. In contrast, rapid regional cooperation in Central Asia took place among unsettled and developing states. How can we explain this?

Induced Cooperation

Between coerced and voluntary solutions are induced solutions. In these cases, a third party acts as a mediator in the bargaining process. It may be a superpower, an international organization, a non-governmental organization, an individual, or a multilateral funding institution such as the World Bank. There have been instances in which third parties have successfully facilitated cooperation over an international river basin and instances in which they have failed miserably.

The case of international water cooperation most often hailed as a success story is that between India and Pakistan, rival states that have fought several border conflicts since partition in 1947. Here a third party, the World Bank, was able to move the initial post-partition plan to a more stable configuration with the 1960 Indus Water Treaty (Michel 1967). Recognizing that partition had created irreconcilable economic and political divisions, the World Bank put forth a proposal in 1954 that gave Pakistan exclusive control over the three western rivers and India exclusive control over the three eastern ones. The World Bank could implement this proposal because it was technically feasible to separate the tributaries of the Indus water system into two distinct systems; at the same time, the World Bank offered side payments to India and Pakistan to reward them for restructuring their irrigation networks. In short, the World Bank helped maximize the benefits to both parties while lessening their dependence on each other, rather than trying to foster a water management arrangement based on mutual interdependence.

For numerous reasons, other cases of third-party intervention at the international level have not been as successful. Lowi (1993a) argues that the 1955 Johnston Mission talks on the Jordan basin broke down because the political rivalries between the Arab states and Israel overshadowed any attempts to first settle the riparian dispute, even with third-party intervention. Similarly, since the Oslo Accords of 1993 third parties have been unsuccessful in reaching a separate water agreement outside of the final status talks. Third parties can exercise influence by conditioning their funding for water development projects in international river basins on the prior negotiation of an international agreement between the affected parties; yet, when a state's interests are inimical to other states, it may choose to undertake a project alone, without outside assistance (as Turkey has done with the Southeast Anatolia project).

Under what conditions do third parties play a role, and to what extent are they able to induce cooperation among states that might not recognize the mutual benefits to be in their individual interest? As already noted, different water systems possess different distributions of capabilities and interests among co-riparians. Coerced cooperation usually appears where there are stark asymmetries among the actors, whereas voluntary cooperation often emerges in situations of symmetrical interests and homogeneity. Yet, as in the Aral basin, there may also be situations where offsetting asymmetries of interests and capabilities define the relationship among the co-riparians. (See table 2.2.) For example, a politically more powerful nation (e.g., Uzbekistan) may be in a weaker position in regard to the physical control of the resource. In cases where there are no stark asymmetries or instances of clear symmetries, depending on how the asymmetries are structured, there may be opportunities for tradeoffs, issue linkages, or a role for a third party.

Where the interests and the capabilities of riparian states are asymmetrical and shifting, an active role for third parties helps to offset the

Table 2.2
Patterns of cooperation in international river basins.

	Interests and capabilities	Pattern of cooperation	Representative real-world cases
Stark asymmetry between upstream and downstream states	Upstream states more powerful economically, militarily, and physically, and more interested in water use.	Coerced cooperation (hegemon)	Euphrates basin (Turkey)
	Downstream states more powerful economically and militarily, and more interested in water use.		Nile basin (Egypt)
Symmetry between upstream and downstream states	In spite of the physical asymmetry between upstream and downstream states, they have similar interests and/or military and economic capabilities.	Voluntary cooperation in form of institutions or compensation	Rhine basin, Columbia basin
Offsetting asymmetries between upstream and downstream states	Upstream state is physically more powerful, but is weaker militarily and economically. Downstream state has great interest in water use and is more powerful militarily and economically.	Induced cooperation via side payments, issue linkages, and/or role for third parties.	Aral Sea basin, Indus basin, lower Mekong basin

asymmetries with the introduction of side payments. Side payments remunerate those who might not gain from cooperation. These side payments or financial and material transfers are crucial for the Central Asian states, which must build the empirical components of statehood while dealing with the problems of managing a shared water system. Moreover, international organizations, bilateral aid organizations, and non-governmental organizations furnish side payments in the form of financial and material assistance to re-equalize the situation and re-create mutual interdependence after state breakup. It should not be overlooked that third-party actors may also provide side payments designed to decrease mutual interdependence after state breakup, as in India and Pakistan.

In the field of international environmental politics, a precedent has already been established for using side payments as a means of bringing about cooperation among developing countries (Hurrell and Kingsbury 1992, p. 23). International environmental agreements that incorporate side payments may be able to manage global common property resources better than ones that preclude side payments (Barrett 1990). Poor countries might not have signed the World Heritage Convention had it not been for the inclusion of side payments, established by the World Heritage Fund. To get the developing countries to sign the Montreal Protocol, the advanced industrialized countries agreed to create the Montreal Protocol Fund, which aided in the transfer of new non-ozone-depleting technologies (DeSombre and Kauffman 1996). When side payments are offered, developing or weakly institutionalized states are likely to forgo sovereignty-enhancing positions and adjust their policy decisions to those of the other actors, sharing a common resource.

In the next chapter I address two-level institution building. My approach to this matter, which relies on third-party actors and side payments, reveals why the Central Asian states were able to overcome the obstacles to collective action in regard to a large common-pool resource immediately after the breakup of the Soviet Union. Had international organizations, non-governmental organizations, and bilateral aid organizations not intervened in Central Asian state building, rapid regional water cooperation might not have occurred.

3

Building Environmental Cooperation under Conditions of Transformation

During the Soviet period, the Central Asian republics did not worry about who had clear legal rights to the use of the fresh-water resources in the Aral basin, because the water system was considered a purely domestic resource within the territorial borders of the Soviet Union. At that time, the Soviet authorities managed the Aral basin as an integrated and highly interdependent system. The breakup of the Soviet Union, however, led to a situation in which the political borders no longer coincided with the physical borders of the principal water system undergirding the economic and social structure of the region. New territorial borders artificially divided up the water system among five states, creating new asymmetries of capabilities between upstream and downstream users and at the same time generating competing interests over water allocations. State breakup, in short, introduced new claims of ownership rights over the transboundary water resources and related infrastructure in Central Asia.

The Central Asian successor states inherited the previous system for water management without a centralized authority to guide it. Accordingly, they could no longer rely on Moscow to allocate water among them and to provide mechanisms for conflict resolution. To complicate matters, the Central Asian successor states were left on their own to address one of the world's largest anthropogenic environmental disasters: the desiccation of the Aral Sea. Before independence, they could blame Moscow and demand compensation from the center to mitigate the devastating environmental and health effects. After independence, Russia and the newly demarcated upstream states renounced responsibility for the water mismanagement strategies that had caused the crisis.

With the demise of the Soviet center, the successor states could have chosen three different outcomes: to abrogate past practices and pursue independent decision making, to maintain previous institutions and continue as before, or to renegotiate new institutions for joint management. The Central Asian successor states chose the third option. Insofar as the benefits of cooperation were asymmetrical and unevenly distributed (because of the physical nature of an international river basin), these newly independent states found that they were unable to proceed alone. As a result, they turned to a broad array of actors from the international community to assist them in building new institutions for water cooperation in the Aral basin.

This chapter explicates why the Central Asian states sought immediate international assistance to deal with a regional environmental problem. In order to explain why members of the international community could assume a catalytic role in shaping the nature and form of water institutions in the Aral basin, I argue that the environmental crisis faced by the Central Asian states must be situated within the context of state formation. As part of state making, a wide array of state and non-state actors strategize to promote institutional change among societies emerging from state breakup and among societies undergoing regime change.

This chapter specifically seeks to explain third-party intervention in the Aral basin; however, its broader purpose is to investigate the pronounced role that international organizations (IOs), non-governmental organizations (NGOs), and bilateral aid organizations have assumed in world politics in the post-Cold War era.[1] Especially since the collapse of the Soviet Union, transnational actors have intervened in the internal politics of post-communist states in a broad range of issue areas, including democracy building, environmental protection, poverty amelioration, and prevention of ethnic conflict.[2] On the one hand, IOs, bilateral aid organizations, and NGOs assist transitional states with state making at the domestic level. These actors play a crucial role in building the institutional foundation of these transitional states by transferring know-how, ideas, and skills concerning such issues as constitutional choice, electoral rules, taxation and budgetary systems, and military and police reform. On the other hand, transnational actors seek to promote regional cooperation among transitional states as a means of mitigating regional instability and integrating new states into the international community of nation-states through such organizations as the North Atlantic Treaty Organization (NATO) and the Organization for Security and Cooperation in Europe (OSCE).

Yet, in view of the sovereignty concerns associated with their newfound independence, why would the Central Asian states ever consent to immediate intervention from the international community to assist them in finding the most appropriate solution to the Aral Sea crisis? In order to understand the significance of sovereignty and sovereignty-enhancing behavior for the newly independent Central Asian states, in the next section I demonstrate why cooperation in the Aral basin was empirically and theoretically puzzling. In the subsequent section, I elucidate why traditional approaches that ignore transnational actors are inadequate for understanding problems of environmental cooperation in transitional situations. In the final section, I amplify the pivotal role of IOs, bilateral aid organizations, and NGOs in creating conditions for interstate cooperation and for state making in Central Asia by means of side payments to domestic actors.

The Puzzle: Cooperation under Conditions of Transformation

State Making versus Regional Cooperation

For the Central Asian states, independence introduced the question of sovereignty, both empirically and legally. Sovereignty is the ability to make one's own policies, or to govern internally within legally sanctioned borders (Thomson 1995). Since new states seek to maximize their sovereignty, I assume that the Central Asian states want to jealously guard their newly acquired sovereignty. In regard to natural resources, this translates into giving priority to independent decision making in place of joint management decisions.[3] Independence endowed each Central Asian successor state with the opportunity to define its own strategy for water use and its own method of control over related infrastructure within inherited republican borders. As expected, one of the first acts of each newly independent Central Asian state was to declare sovereignty over the use of its natural resources. This is not dissimilar from when post-colonial and developing states invoked doctrines such as "permanent sovereignty over natural resources" to assert their independence from colonial powers or from other foreign actors operating within their territory.[4] Laying claim to independent policy making within territorial borders is an essential part of state making.

Proclaiming independence and control over domestic natural resources and economic issues from actors outside of internationally recognized borders establishes "borders of separation" (Kratochwil 1986, p. 42). Although the post-communist states emerged from similar political and economic institutions, independence severed many of these ties, even in a region bound together by cultural, linguistic, and historical factors. Each new state sought to distinguish itself from the others and (most important) from any relics of the Soviet legacy.[5] For that reason, the successor states were especially sensitive to any Russian involvement or encroachment in their domestic and foreign policies. While borders of separation are constituted by the recognition of territorial borders, they also require the consolidation of domestic institutions. Thus, independent decisions concerning water allocation and water appropriation in Central Asia were parts of the broader process of the successor states' establishing markers of their separation from Moscow and from one another.

Though the newly independent Central Asian states were concerned about constituting borders of separation, they nevertheless also had to address the question as to whether they would cooperate to share their water resources or pursue independent strategies that could inflict additional harm on neighboring states and—if they chose to cooperate—the question as to what form the cooperation would take. Despite the alarming predictions of discord over water use, the Central Asian states engaged in cooperative behavior with one another after the breakup of the Soviet Union. Notwithstanding independence and state-making concerns, the removal of an external decision-making authority in Moscow did not precipitate conflict. As early as November 1991, the five ministers of water management of the Central Asian states began to work on the development of cooperative management schemes for the Aral basin. On February 18, 1992, after gaining independence, the five ministers signed the first agreement on "Cooperation in the Management, Utilization, and Protection of Water Resources of Interstate Sources," in which the states "commit[ted] themselves to refrain from any activities within their respective territories which, entailing a deviation from the agreed water shares or bringing about water pollution, are likely to affect the interests of, and cause damage to the co-basin states."

The overarching question that this book seeks to address is this: Why was there interstate environmental cooperation rather than acute conflict in Central Asia immediately after the collapse of the Soviet Union? As I noted in chapter 1, scholars and policy makers assumed that the unsettling of political and physical borders would intensify violent conflict and competition over land and water resources in Central Asia rather than engender the political conditions necessary for cooperation to take hold. To the surprise of many observers, the Central Asian states took the initial necessary steps to introduce new institutions for water management that corresponded to the changed political reality.

The dependent variable is the process of rapid regional institutionalization of water sharing among the five Central Asian states in the period 1992–1998. Since most institution-building efforts involve highly contested negotiations at both the domestic and the international level, I assume institutions are necessary for cooperation.[6] Institutions are social constructs created by individuals or governments to cope with the problems of coordination and cooperation that arise as a result of interdependence.[7] In an international river basin, if actors follow independent policies, their actions may interfere with the efforts of others to pursue their own ends, and in turn, they are each left worse off than if they had chosen to mutually adjust their actions.[8] The type of cooperation through institutions can be both informal and formal, but I adhere to a slightly restricted definition of institutions as the indicator for cooperation similar to "multilateral agreements among states which aim to regulate national actions within an issue-area."[9]

In the Central Asian case, rapid regional environmental cooperation entailed negotiating new interstate agreements to mitigate a problem inherited from the Soviet period and establishing a mechanism for future interaction. Since one cannot understand cooperation without understanding the frequent absence of cooperation, the counterfactual in the case of the Aral basin is non-institutionalization or non-cooperation—

that is, discord.[10] The implicit counterfactual is the ethnic variable. Unlike many of the other Soviet successor states, the Central Asian states avoided acute ethnic conflict immediately after independence.[11] In short, any of the above outcomes would have been utterly sub-optimal for the newly independent states, which faced a severe environmental crisis while lacking the basic empirical institutions for statehood.

By consenting to multilateral arrangements, states abandon the presumption that they can act unilaterally; as a result, cooperation implies that each state must take the interests of others into account before devising its own policies. Because they achieved independence only after the collapse of the Soviet Union, the newly independent Central Asian states should have been reluctant to forgo independent policy making immediately afterward. Moreover, since independence was thrust upon them, the Central Asian states were not well prepared to deal with the tasks of state making and domestic governance in a changed international context (Olcott 1993). Indeed, sovereignty concerns associated with independent policy making should have led to predictions that these states would fail to cooperate over their fresh-water resources in the Aral basin.

Cooperation in Settled versus Unsettled Situations

The emergence of cooperation in the Aral basin so soon after independence was especially striking since most other attempts at rapid regional institutionalization and cross-border exchange within the territory of the Newly Independent States have been futile.[12] In other cases dealing with shared natural resources or infrastructure, the successor states of the Soviet Union preferred to solidify their borders rather than to build new institutions for cooperation. For example, soon after the breakup of the Soviet Union there was tension between Russia and Ukraine over control of the Black Sea fleet. Within the Caspian basin, the littoral states failed to specify its legal demarcation and clashed over pipeline routes and ownership of oil stocks.[13] Even among the Visegrad states in East Central Europe, regional cooperation fell apart once states' interests began to differ (Bunce 1997). And after the breakup of Czechoslovakia, Hungary and Slovakia became immersed in a non-violent dispute over the Gabcikovo-Nagymaros dam project on the Danube that had to be adjudicated by the International Court of Justice.

Even though the achievement of interstate cooperation is considered unusual, it is known to occur under particular conditions, such as where there are long-term horizons, strategies of reciprocity, and common interests.[14] Cooperation is more likely among certain kinds of states and over certain issue areas. Since World War II many scholars have observed a rise in multilateral cooperative arrangements among the advanced industrialized countries and what can be considered settled states (Ruggie 1992). In particular, the growth of multilateral institutions among the advanced industrialized countries of Western Europe has facilitated the coordination of their economic and foreign policies at the intergovernmental level; especially in the economic and environmental spheres, states have realized mutual gains. Thus, the advanced industrialized countries usually interact in an iterated manner.

Insofar as the advanced industrialized democracies are less likely to be prone to conflict, they are better equipped to deal with problems of collective action that arise from interdependence. The subfield of international environmental politics is replete with examples of the successful resolution of collective-action problems. Here, once again, cooperation occurs among settled states and settled situations, as in the Rhine basin. The Montreal Protocol, a benchmark in environmental treaty negotiation, is often hailed as an impressive blueprint for international cooperation in reducing a global threat to human health and the physical environment; the advanced industrialized countries that also were some of the largest producers of the ozone-depleting substances led the push for global controls (Benedick 1991; Haas 1992). The 1979 Long-Range Transboundary Air Pollution Convention also stands out as a positive case, as it succeeded in curtailing acid rain producing emissions (Levy 1993).[15]

In contrast, environmental cooperation in Central Asia took place among unsettled states during a period of domestic and international transformation. This process of rapid regional institutionalization in Central Asia is puzzling for the following reasons.

First, the Central Asian successor states were poor and weakly institutionalized, lacking the technical capacity and financial resources to ameliorate environmental problems. In comparison with interstate environmental cooperation among the advanced industrialized countries, developing countries have a poor record of reaching cooperative environmental agreements. Yet these are the very countries that have the greatest need for environmental cooperation to facilitate economic development. Developing countries are often reluctant to sign onto many of the global environmental accords for fear of relinquishing sovereignty over internal matters. The Central Asian states should have resembled most other developing countries that lack the domestic capability to cooperate with other states on environmental resources.

Second, interstate cooperation is unlikely when one state gains more and can use its gains to threaten another state in future interaction or where the danger of defection by one state cannot be hedged against in the event of unanticipated defection.[16] State breakup and the subsequent shifting of political boundaries in Central Asia introduced a situation of asymmetries of capabilities and interests in the Aral basin, and with the removal of an external enforcer, each new state became especially sensitive to the other's relative gains during the negotiations over the various options for water-sharing arrangements. The initial agreement not only specifies who benefits at this first round but also determines the distribution of benefits for future rounds, and therefore any initial water agreement among the Central Asian successor states could constrain these states in later rounds when and if they chose to amend or revoke the agreement. Since the Central Asian case deals with asymmetries created by the physical properties of a resource and its infrastructure, a greater potential for reneging on any commitment exists. In Central Asia, the original Soviet conditions conferred most of the benefits to the downstream states (Kazakhstan, Turkmenistan, and Uzbekistan), which produced the bulk of the Soviet cotton crop. But after independence the downstream states were unable to prevent the upstream states (Kyrgyzstan and Tajikistan) from failing to abide by customary patterns of water sharing since the upstream states inherited most of the reservoirs and dams.

Third, rapid regional cooperation in Central Asia is puzzling precisely because it occurred *under conditions of transformation.* The literature on political transitions away from authoritarianism toward democracy points out that the period of extrication from the old regime until the time of consolidation of the new regime is characterized by uncertainty.

The rules of the game are not clear (O'Donnell and Schmitter 1986). This raises similar defection concerns, as mentioned above, since the new rules are yet to be clearly specified. If anything, the rules are constantly in flux as new leaders are trying to stake out their claims and only focus on short-term gains. In most of the post-communist cases, long-term horizons are absent and enforcement mechanisms are tenuous, which in turn encourages sovereignty-enhancing positions rather than cooperative arrangements from emerging so soon. In addition, some scholars find that countries undergoing the process of democratization are much more conflict-prone (Mansfield and Snyder 1995). In fact, the end of the Cold War has seen a rise in conflicts regarding transitional states, including conflicts between Serbia and Croatia and between Armenia and Azerbaijan.

Most important, the Central Asian successor states were not just engaged in political transitions. Like other post-communist states, they needed to undertake revolutionary political, economic, and social change.[17] Cooperation and the creation of institutions are contentious processes even for settled states; they should be even more difficult and complicated for transitional states simultaneously engaged in institution building. The successor states of the Soviet Union and East Central Europe needed simultaneously to construct domestic state institutions and to constitute new interstate institutions. Contrary to previous periods of state formation in Western Europe and Africa, the unsettling of the borders in the Soviet Union and Eastern Europe required these post-communist states to reconfigure the basic constitution of the nation-state while concurrently restructuring their domestic political and economic institutions away from state socialism so that they could integrate into the global liberal economic order.[18] All in all, post-communist state formation does not proceed from domestic struggles between state elites and society, from military interactions with other states, or solely through juridical recognition; rather, as the next section shows, state formation in the post-Cold War context is embedded within a complex and varied network of transnational relations among state and non-state actors. In sum: Owing to the magnitude of the transition from state socialism, we should not have expected environmental cooperation immediately in Central Asia.

Integrating International and Domestic Politics

The above puzzle presented by regional environmental cooperation among the newly independent Central Asian states demands an integrative approach that connects domestic and international level politics. Yet conventional theories of interstate cooperation often maintain a stark separation between domestic and international processes. In the field of international relations, both realist and institutionalist theories (and also what are commonly referred to as neorealism and neoliberal institutionalism) assume the primacy of the state and conceive of it as a unitary, rational actor.[19] They are similar in that they locate the causes at the systemic level, whereas realist explanations and to a lesser extent institutionalist explanations maintain that the rise of cooperation depends largely on a set of constraining conditions: the structure of the international system and the distribution of power within it. At the same time, others suggest that these approaches are insufficient for explaining cooperation and can only provide a first cut; instead, these scholars suggest an alternative explanation that draws attention to domestic approaches or the second image and posit that the impetus for cooperation emanates from within states rather than from the overarching structure in which states find themselves (Gourevitch 1978). In short, this standard debate regarding why states cooperate only reinforces the dichotomy between international and domestic processes.

Although there is an extensive literature on state formation and a well-developed literature on international cooperation, few have sought to integrate the two in order to link these domestic and international processes.[20] Among the few efforts to bridge this gap between domestic politics and international relations, Robert Putnam's work on two-level games stands out. Putnam (1988) links the two levels by modeling international negotiations as a two-level game wherein he shows how a chief negotiator or leader will find it necessary to first reach an agreement with other governments and then secure domestic ratification. In these circumstances, a negotiator often uses the threat of a domestic veto in the subsequent ratification process as means to tie his or her hands in the current negotiations.[21] A fragmented domestic environment in which numerous interest groups compete for influence with legislators can hinder or even preclude an international agreement from ever being implemented. Since these two-level games are largely models of international bargaining situations, emphasis is placed on the achievement of mutual gains at the international level; the domestic situation instead serves as an instrument that can be employed at the international level in order to obtain these gains. As a result, these approaches based on two-level games often understate the domestic gains that can be accrued through international cooperation. International bargains are also about the distribution of costs and benefits among domestic groups, but this requires going beyond just viewing the domestic realm as a constraint (Evans 1993).

Although the integration of domestic and international factors helps to explain the rapid regional cooperation in the Aral basin, it is nevertheless essential to evaluate how useful these conventional approaches based on two-level games are for expounding bargaining situations in post-communist states, insofar as these are still states in the making. In like manner, Evans, Jacobson, and Putnam (1993, p. ix) ask to what extent Putnam's insights and generalizations are applicable to negotiations in non-Western countries and to other non-economic matters. Since the case of Central Asian water clearly involves non-Western countries and a scarce environmental resource, it provides an opportunity to extend and enhance these approaches based on two-level games. In short, the application of a refined two-level game to post-Soviet Central Asia reveals that IOs, bilateral aid organizations, and NGOs are assuming major roles as mediators between domestic politics and interstate relations.

Approaches based on two-level games challenge the assumption that one can model the state as a unitary actor by incorporating various domestic interest groups as constraints on the ratification and implementation phases, but they exclude other actors that are not constituents of "the state" in the actual bargaining game. Thus, how relevant are they for capturing bargaining situations in unsettled situations and among unsettled states where the boundaries are ambiguous both geographically and institutionally? Precisely because the political boundaries have shifted and domestic institutions are inchoate, traditional two-level-game approaches require a more nuanced version that adds a third level to the game: the level of transnational actors. In general, third-party or transnational actors facilitate negotiations by providing information and lowering the transaction costs. IOs, bilateral aid organizations, and NGOs assume a much more active and even proactive role in inducing cooperation among new states that have emerged from state breakup. Through their links to actors within the state and to other non-state actors, transnational actors influence policy changes and institutional development. In short, they act as agents of change within the realm of domestic politics and within the international system.

Thus, what distinguishes environmental cooperation among post-communist states is the internationalization of the process.[22] Bargaining among transitional states involves purposive third-party actors with their own unique well-ordered preferences, which are distinct from those of any of the concerned governments. With few exceptions, regional environmental-protection programs are embedded within the international system in which transnational actors interact with states but can also bypass the states to influence important domestic actors that can put pressure on a state's policy choices. Once traditional two-level-game approaches are refined to include transnational actors, they are able to explain regional cooperation among transitional states and to illuminate the complex and cumbersome process of post-Cold War state making.

New leaders, moreover, seek to participate in international negotiations and to join international organizations as a means to consolidate their domestic power base and to gain domestic legitimacy both at home and abroad. Taking this one step further, the process itself of engaging in international negotiations for regional cooperation strengthens the state rather than undermines it.[23] For many weak states, membership in the United Nations is the only way to ensure independence and sovereignty (Jackson 1990). In fact, sovereignty concerns do not preclude cooperation. Indeed, the reverse is possible: Owing to the prestige that is associated with becoming a member in the international community of nation-states, cooperation may strengthen state sovereignty. Engaging in foreign relations with neighboring states legitimates a state's territorial borders through mutual recognition. Similarly, when states join international organizations (e.g., the United Nations, the Organization for Security and Cooperation in Europe, and the World Trade Organization), they are, in effect, agreeing to adopt certain norms and ideas about how states should behave in the international system and domestically. For example, the OSCE considers the protection of human rights to be of the highest priority and thus expects its member states to respect human rights domestically. For the Central Asian states, this has meant that they have had to address the demands and concerns of the Russian minority populations in order to be considered members in good standing (Olcott 1996, p. 174).

Thus, where the juridical and empirical components of statehood are just taking shape, as in Central Asia, the process of designing new institutions for interstate exchange may be just as much about state making as it is about promoting regional cooperation. In the following section, I use my approach to two-level institution building to explain the rapid emergence of regional environmental cooperation in Central Asia. By taking into account the porosity of boundaries territorially and institutionally and the flux of the transition period (which engenders a dynamic role for transnational actors), I integrate theories of domestic and international politics.

Two-Level Institution Building

The breakup of the Soviet Union and the subsequent end of the Cold War presented an opportunity to build on the work of scholars who had sought to integrate domestic and international politics. In the words of Keohane (1993, p. 294), there is a need to "link domestic politics with international relations in a theoretically meaningful and analytically way." Any theoretical study of cooperation and institution building in the post Cold War period requires that the boundaries between international and domestic politics be erased, since empirically the interrelated processes of state breakup and state formation call into question traditional understandings of territoriality and practices of national sovereignty. First, states are no longer the only major actors in the international system, and more attention should be paid to the positive and negative influences of transnational actors.[24] Second, political territorial boundaries are no longer sufficient to differentiate international and domestic politics now that IOs, bilateral aid organizations, and NGOs are assuming many of the functions of domestic governance (especially in transitional and developing countries).[25]

Regional cooperation under conditions of transformation occurs at the intersection of changes in international and domestic boundaries.[26] My alternative approach to environmental cooperation among the Central Asian states is predicated on finding the nexus of international and domestic explanations. Bridging comparative politics and international relations theory, it specifies how the nature of the multifaceted domestic transitions away from state socialism affects the prospects for interstate cooperation while also reversing the causal flow to grasp how the building of institutions for regional cooperation shapes the overall state-building process. The three elements of two-level institution building are *simultaneous institution building, third-party actors,* and *side payments.*

In a transitional context, the duality of institution building engenders a situation in which leaders explicitly turn to the international setting in order to achieve domestic aims. My approach to institution building differs from the conventional two-level-game approach, which suggests that states are more concerned about achieving international gains than domestic ones. In Central Asia, negotiations over interstate cooperation bolster the internal position of the new leaders. The primary goal of those leaders is to stay in power, which requires appeasing important domestic constituencies. Most important, I argue, the bargaining over new water institutions is being driven mainly by third-party actors—IOs, bilateral aid organizations, NGOs—rather than by national leaderships.

Simultaneous Institution Building

The first element of my approach consists of treating international and domestic processes in transitional states as simultaneous. Rather than consider the negotiations over the formation of international institutions as the primary game, equal weight is given to the domestic context wherein a second bargaining game is unfolding over the design of the future political and economic institutions that make up the state. The newly independent Central Asian states faced two types of collective-action problems as a result of the collapse of the Soviet Union. At the international level, they needed to figure out cooperative ways to share their joint water resources and to mitigate environmental externalities; at the domestic level, they needed to undertake three simultaneous transformations—political, economic, and national. The underlying assumption is that most transitional states must move away from state socialism and toward capitalist liberal democracy.[27]

By focusing on the interaction between interstate cooperation and domestic state building, I argue that regional environmental cooperation in Central Asia can be explained only by nesting it within the process of state building. Broadly, regional cooperation is a subset of the larger game of building states from without before having to build them from within. States are legitimated through international recognition even before they possess the internal capacity for decision making (Jackson 1990). Leaders exploit the international aspects of legal sovereignty to garner domestic legitimacy in place of having to concentrate on establishing the empirical components of sovereignty. Merely coordinating activities with IOs and sending delegates to international conferences give the newly independent Central Asian states a sense of "stateness."

Thus, what appears to be a matter of interstate cooperation is actually a matter of multiple bargaining games in multiple arenas at the domestic and international levels. While state leaders are bargaining over water, they are also bargaining with different economic sectors with vested interests in the water sector and bargaining over the form and scope of domestic institutions that are necessary to regulate water sharing among and within states. The driving force underlying this argument is that domestic and international processes are mutually reinforcing and constitutive, rather than contradictory or in opposition to one another. It is the manifold transitions away from state socialism that led the newly independent Central Asian states to engage immediately in cooperative efforts to redesign institutional arrangements for interstate water sharing. In fact, bargaining over the formation of new international institutions entails resolving both types of collective-action problems simultaneously, which forces leaders to negotiate concurrently at two tables. At the interstate level, a leader must bargain with other leaders over whether to devise an institutional agreement and if so then over the specific content of the agreement. At the domestic level, a leader of a transitional state needs to ensure compliance with the interstate agreement by conducting parallel negotiations with critical domestic constituencies while also seeking to negotiate the design of the new domestic institutional structures of the state apparatus. Although these two sets of institutions are distinct, their negotiations are inevitably intertwined because effective states are necessary for carrying out interstate agreements.

Third-Party Actors

The second main component is the inclusion of third-party or transnational actors in the simultaneous bargaining. IOs, bilateral aid organizations, and NGOs have a crucial role in creating both the conditions for regional cooperation and the conditions for state building. In fact, recent research has focused on the manner in which NGOs and IOs, through transnational linkage and financial assistance, are able to shape domestic and foreign policies in both developing and developed countries.[28] Pressure from local NGOs and grassroots movements has forced some major international lending organizations, including the World Bank, to reconsider many of their development programs that might cause environmental harm or human displacements in the countries in which they operate (Fox and Brown 1998). NGOs, in particular, assume a crucial role in transferring social norms, framing issues and generating and organizing information (Keck and Sikkink 1998). IOs contribute to improving institutional effectiveness by increasing governmental concern, enhancing the contractual environment and building national capacity (Haas, Keohane, and Levy 1993). In similar manner, IOs and NGOs have been instrumental in influencing state behavior in Central Asia. Along with bilateral aid organizations, they have helped to define the agenda, choose the participants, and construct the alternative negotiating sets; thus, they have influenced the form and the scope of the institutions that have emerged in the Aral basin. To incorporate actors that transcend boundaries in the post-Cold War context, the bargaining space must be broadened. Unlike other developing countries that emerged onto the international system of nation-states after the period of decolonization in the 1950s and the 1960s, post-communist states have unquestionably accepted a wide role for IOs, bilateral aid organizations, and NGOs in building institutions at both the interstate and the domestic level.

Such an enlarged role for third-party actors in resolving collective-action problems contrasts with the realist and institutionalist approaches for understanding cooperation. Contrary to realist expectations, a hegemon is unable to coerce cooperation under conditions of transformation.

Contrary to institutionalist approaches, transitional states do not voluntarily come to the bargaining table to negotiate new interstate agreements. Instead, third party actors induce cooperation among transitional states by negotiating with the dominant players (i.e., state elites) at the international level while simultaneously bargaining with a wide array of societal interest groups at the domestic level.

Furthermore, realist, institutionalist, and even traditional approaches based on two-level games usually focus on the state as the central actor and exclude outside actors that might challenge the permeability of the state as an actor. My approach to two-level institution building emphasizes the role of transnational actors such as IOs and NGOs and moreover regards them as autonomous and with interests.[29] By considering international organizations as autonomous, my approach does not conflate or reify institutions with organizations, as was characteristic of the earlier international organizations literature whereby international institutions were coterminous with formal organizations.[30] Not turning them into actors such as organizations allows us to look at the relationships among institutions, organizations, and states (Rosenau 1986, p. 882). It is essential to approach them separately because there is a growing legitimacy of international involvement in the internal politics of states, especially transitional states. At the international level, some scholars detect that more and more IOs are creating the institutions for cooperation and international governance. For example, Young (1994a, p. 105) points out that while "not suggesting that states no longer dominate bargaining in international society, it would be a serious mistake to overlook the role of transnational alliances among influential interest groups in developing and maintaining governance systems at the international level."

In addition, unlike early international relations theories that saw the rise of transnational actors or even supranational organizations as a reflection of a growing integration and supersession of the nation state, the enlarged role of international organizations can uphold and bolster the institution of the nation-state, albeit calling into question traditional meanings of state sovereignty. Ironically, non-state actors often violate state sovereignty to maintain state sovereignty.[31] Thus, my approach is not a return to a functionalist or neofunctionalist argument that would envisage specialized international agencies and their technical experts or supranational bureaucracies as an indicator of regional integration or the decline of the nation-state system.[32]

Side Payments

Finally, I connect domestic politics with international politics through side payments,[33] which substitute for the lack of resources needed by transitional states. This is a point of departure from the existing literature on cooperation that essentially regards the use of side payments as primarily a bargaining tactic at the domestic level to gain domestic ratification (Putnam 1988, p. 450). With respect to the use of side payments, Milner (1992, p. 473) argues that much of the cooperation literature (even work based on two-level games) has failed to specify "when and how side payments are made." By focusing on the Aral basin, I elucidate precisely when and how side payments induce institution building in transitional states at both the interstate level and the domestic level.

In order to then capture the multiple institution-building processes at both the interstate and the domestic level, interstate cooperation under conditions of domestic transformation is conceptualized as a two-level interaction in which *the need for* side payments at the domestic level and *the willingness to* provide these at the international level forms the nexus between the two levels. Unlike third-party actors in the Cold War period, IOs and NGOs in the post-Cold War period occupy a pivotal role in creating both conditions for interstate cooperation and conditions for state building in transitional states through side payments. In fact, such transnational actors abet institution building at both the international and the domestic level.

Linking the Two Levels: The Role of Third-Party Actors and Side Payments

In this section, to demonstrate how two-level interaction operates, I depict the salient forces at both the international and the domestic level. At the domestic level, I highlight the internal nature of transitional states.[34] At the international level, I closely examine the role of third-party actors, focusing on transnational actors. The interaction of these two factors explains both interstate cooperation and the particular type of state building in transitional states generally and in Central Asia specifically, which occurs through the transfer of side payments from transnational actors to domestic interest groups.

The Domestic Level: Weakly Institutionalized States

At the domestic level, the type of state matters especially since states are involved in negotiating and ultimately implementing new institutions for cooperation. The breakup of the Soviet Union shifted and reconstituted political borders, creating fifteen successor states. Unique and common to all the Soviet successor cases is that in addition to the unsettling of borders, their internal domestic situation was altogether destabilized. In the aftermath of the dissolution of the Soviet Union, the successor states had to embark on a process of state formation that was unlike early state formation in Western Europe or that which resulted from the period of decolonization in the 1960s. Post-communist state building is embedded in the particular historical circumstances of the post-Cold War era, identified by a liberal economic order. The big multilateral lending organizations such as the International Monetary Fund and the World Bank, along with bilateral aid organizations such as US Agency for Information and Development (USAID), are the driving forces behind many of the policies associated with promoting this liberal economic order.

Central Asian state building is a subset of post-communist state building. When the Soviet Union collapsed, all the former Soviet republics and many of the Eastern European successor states faced unprecedented economic, political, and social transformations. This similar class of problems arising from the process of state breakup is the "initial conditions" in which state leaders make crucial choices concerning how to build new domestic political and economic institutions and new cultural and social identities. In the political arena, post-communist states are undergoing a regime transition that requires opening up the political arena to contestation. At the same time, these newly independent states must restructure a highly inefficient economic system where markets were absent; instead the economy was characterized by soft-budget constraints, regional specialization, and centralized planning to meet set production targets.[35] Finally, these successor states must foster new national identities as part of the process of nation building, resulting in the resurrection of past cultural heroes, the rewriting of history, and the introduction of new language and citizenship laws.

Transitions, by definition, are about change. Not surprisingly, the multifaceted transitions away from state socialism entail a period of radical restructuring, and any study of the political behavior of transitional states must view them in a dynamic setting. The endpoint of these transitions remains ambiguous, which only heightens the uncertainty of the transition. In fact, uncertainty is one of the major factors characterizing these domestic transformations; it underlies domestic institution building, especially since these states had no historical archetype of other multiple transitions upon which to draw and instead must consciously renegotiate new institutions that will structure the future internal relations between state and society and define the mode of domestic governance. Simply put, the post-communist transitions differ from other types of transitions precisely because they are about making states (Linz and Stepan 1996).

Yet it is also important to note that change is conditioned by how a country extricates itself from the previous system. The post-communist transformations are not about revolutions, although the change is clearly revolutionary. They are also not about the collapse of an empire, although disparities existed between the periphery and the center in the Soviet Union. Rather, post-communist state formation is a result of state breakup, the collapse of a multi-ethnic state, and the subsequent reconstitution of borders.[36] In the Central Asian republics, struggles for independence were largely absent; the rupture with the past came unexpectedly, catching the populations ill-prepared for independence. The legacy of 70 years of Soviet rule does not fade quickly; it lingers and undergirds the process of state formation, permeating all aspects of domestic institution building.

In the Central Asian situation, it is necessary to understand the implications of state breakup in order to comprehend the full nature of the transitional period. Most important, the lack of an independence struggle meant that these new states lost both their main source of resources and the mediator of inter-republic conflicts when the center disappeared. This created a void in both authority and in rents that needed to be filled. Without Moscow, the newly independent states confronted the tenuousness of the multifaceted transition without the political authority to mediate conflicts or the financial and material resources to continue to placate important domestic constituencies. Members of the water nomenklatura found themselves cut off from the stream of subsidies and resources transfers that they received from Moscow for maintaining the extensive irrigation networks geared toward cotton production.

Although the Central Asian states have numerous similarities to the other post-communist states, they also resemble the African quasi-states (owing to the particular historical circumstance that situates them within the context of the international system).[37] Quasi-states are first and foremost defined by their territoriality, and in Central Asia (as in the post-colonial African states) statehood was consolidated through juridical means even before the empirical elements of statehood were established. For quasi-states, sovereignty is linked to the international community by which membership in the United Nations enfranchises these states through recognition of their independence. In Central Asia, juridical sovereignty corresponded in tandem to the inherited territorial administrative delineations of the Soviet republican borders that were transferred with independence.

Yet, at the same time, these post-communist states differ from the post-colonial states and quasi-states of the 1960s because they entered an international system dominated by a liberal economic order in the 1990s. The breakup of the Soviet Union made obsolete the main political, economic, and ideological alternative to the Western system of governance in the twentieth century (Fukuyama 1989). Rather than encounter an international system characterized by bipolarity and competition between the superpowers, the Central Asian successor states found a fundamentally changed international context in which they were forced to espouse the terminology of democracy and markets in order to integrate into the system. These norms associated with the liberal economic order were embraced before conditions that would support the building of new domestic institutions associated with democratic governance and the free market were in place.

Post-colonial states have often viewed the international community as an extension of their dependence on the former colonial powers, yet here again the Central Asian states differed from the other quasi-states because they did not view the international community as a pariah. Instead the newly independent Central Asian states saw the international community as a way to extract themselves from the legacy of the Soviet Union. As a result, they willingly made both concrete and symbolic overtures to demands from the international community to democratize and marketize to varying degrees even when they would want to retain the past system of political and economic governance for domestic factors related to social and political control. For example, Kyrgyzstan was the first Central Asian country to enthusiastically adopt an IMF sponsored economic reform package.

In summary: At the domestic level the Central Asian transitions combine the specifics of post-Soviet state building with the particular historical context into which they emerged. The fact that these were weakly institutionalized states had direct implications for how these states interacted with other state and non-state actors in the international system. Being weakly institutionalized and transitional created an unforeseen opportunity for transnational actors to intervene and assume a catalytic role. Overall, the position of weakly institutionalized states in the international system and the transformative nature of the internal political and economic institutions influence the extent to which IOs, bilateral aid organizations, and NGOs induce interstate cooperation among them.

The International Level: An Enhanced Role for Third-Party Actors

The preceding subsection emphasized that not all states are functionally equivalent units in the international system. This subsection emphasizes that they are not the only actors. In addition to examining the internal composition of transitional states and their position in the international system, it is essential to study the role of other transnational actors that transcend and cross state borders. Both realist and institutionalist theories fail to consider these actors as autonomous and purposive, even when taking into account the growing economic and physical interdependence of the international system. Yet IOs and NGOs are assuming an expanded role in building institutions for interstate cooperation and in building domestic political and economic institutions. With respect to transitional states, transnational actors negotiate directly with both government elites (e.g., republican and regional leaders) and non-government elites (e.g., domestic constituencies such as farmers) to bring about interstate cooperation and to strengthen the internal domestic capacity of states. In many instances, their bargaining positions represent their organizational interests and are supported by the financial and material resources that they bring to the table—resources that many developing and new states lack and need.

In the case of institution building at the international level, IOs and NGOs energize and fortify a bargaining process to establish interstate institutions for cooperation. In contrast with the Cold War period, during which the rivalry between the superpowers eclipsed the role of third-party actors, the post-Cold War international system is dominated by multiple actors that include IOs such as various United Nations bodies, financial organizations such as the IMF and the World Bank, bilateral and multilateral aid programs such as USAID and the European Union's Technical Assistance for the Commonwealth of Independent States (EU-TACIS) along with a plethora of NGOs covering human rights, labor, and the environment.

For transitional states, IOs and NGOs are instrumental in coordinating and shaping the framework of negotiations among the different actors with a stake in devising new institutional arrangements. An increasingly visible role for IOs and NGOs led Young (1994a, p. 170) to note that in the field of international environmental politics "IOs have become a source of leadership in environmental negotiations, a development that makes it appropriate to speak of them as architects of the institutional arrangements emerging from these negotiations." Some organizations, including the United Nations Environmental Program (UNEP), preserve a highly apolitical role in international environmental negotiations; other organizations, such as the IMF, are embroiled in heated negotiations over the implementation of structural readjustment programs.[38]

The inclusion of transnational actors involves more players in the bargaining, which challenges the authority and the territorial integrity of the nation-state (Litfin 1993, pp. 94–118).[39] The impact of third-party actors is not restricted to the environmental sphere. Their diffuse role is also quite evident within the post-communist states. IOs and NGOs have made concerted efforts to promote regional cooperation and to prevent acute conflict among and within the successor states of the Soviet Union and in East Central Europe (Chayes and Chayes 1996; Mendelson and

Glenn 2000). The OSCE has sought to inculcate a respect for human rights; NATO has sought to transform the militaries of East Central Europe and Central Asia into proponents of democracy by incorporating them into the Partnership for Peace Program (Perry 1996). By conceptualizing IOs and NGOs as separate and autonomous actors, we can assess their specific effects on the creation or redesign of institutions for collective action among new states.

At the domestic level, IOs and NGOs also take on an enlarged role in shaping the process of domestic institution building—that is, they actively seek to assist these transitional states gain the capability for internal policy making. Customary notions of state sovereignty are disregarded in order to address the needs of local populations that domestic governments are unable to meet. In many cases, IOs and NGOs assume the responsibilities of domestic governance and hence negate attempts at independent decision making in the domestic realm. Interventions range from sending in peacekeeping missions, providing humanitarian assistance, and assisting with economic development programs (Helman and Ratner 1992–93). Precisely because the Central Asian states were weakly institutionalized states, transnational actors had greater access to bypass the state and participate in the direct negotiations over new institutions for environmental cooperation and at the same time engage government and non-government elites over how to restructure the empirical components of the state so that they acquire at least superficially the "basic elements" of statehood.

IOs, in particular, abet post-communist state building in two main ways. First, by recognizing their legal sovereignty, the international community grants these states, especially the political leadership, not only international but also a newfound domestic legitimacy.[40] Second, by providing financial assistance for domestic economic and political reforms, IOs, bilateral aid organizations, and NGOs actively promote the growth of democracy and markets in the post-communist states. Western donors consider NGOs to be the building blocks of civil society and as a result, they have sought to support an independent sector not tied to the state.[41] As part of their international democracy-building efforts, organizations such as the National Democratic Institute for International Affairs and the International Republican Institute were geared toward supporting the formation of political parties and running elections in Central Asia. Because these newly independent Central Asian states lacked the financial resources and internal capacity to provide socio-economic goods for their populations, international actors filled this void of empirical sovereignty, which in turn increased their influence in shaping the overall domestic state-making process.

In summary: The active role of the international community in the form of IOs and NGOs impels institution building at both the international and the domestic level by simultaneously inducing cooperation and reinforcing empirical sovereignty. If these transnational actors did not behave purposively and assume such a comprehensive role, other outcomes might have transpired such as inertia, a different form of state building, or a lack of cooperation.

The Missing Link: Side Payments as Compensation

The missing link between the domestic and international levels is how transnational actors dispense side payments. In theory, third-party actors offer side payments to governments as a means to induce interstate cooperation; in practice, they are actually used at the domestic level as a form of compensation to regional constituencies that are negatively affected by the transition away from state socialism. Ironically, it is this active role of IOs and NGOs in providing side payments in the form of financial and material assistance to important domestic constituencies in the transition period that enables interstate cooperation to evolve. Here, the intersection between domestic and international processes and the porosity of borders becomes clearly visible since the internal nature of weakly institutionalized states at the domestic level creates a need for side payments, which IOs, aid organizations, and NGOs, with their enlarged role in world politics, are eager to supply. Figure 3.1 illustrates how the use of side payments brings about institution building at both the international and the domestic level.

In bargaining theory, side payments are simply the extra payoffs to get actors to reach an agreement that are not part of the agreement itself (Schelling 1960; Raiffa 1982). They can be paid to the actors sitting at the table or to actors not sitting at the table who can potentially undermine or interfere with the agreement being made at the table. In transitional states

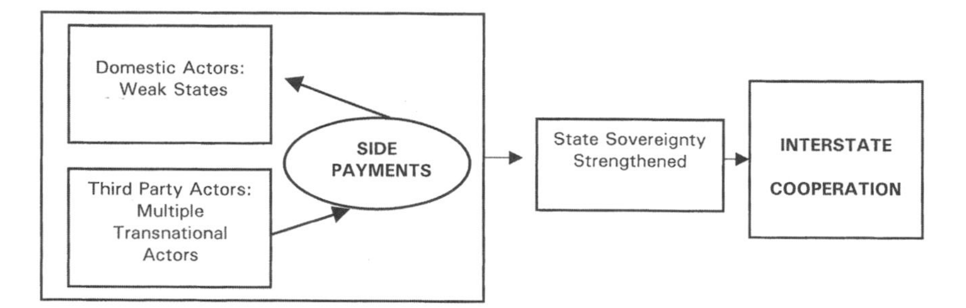

Figure 3.1
Two-level institution building.

such as in Central Asia, IOs such as the World Bank or bilateral assistance programs such as USAID must fund and target a broad spectrum of domestic actors to persuade the government leaders to reach and support an international agreement governing water use.

Even though third-party actors intervened in Central Asia immediately after the Soviet Union's collapse with the stated objective to foster interstate water cooperation, they became entangled in the domestic game of state building. Indeed, third-party actors were actively involved in a simultaneous two-level bargaining game: At one table, they needed to negotiate with the Central Asian governments to reach an international agreement over water sharing; at the other table, they needed to negotiate with the local communities hardest hit by the transitional period to ensure compliance with the agreement. Bargaining over the formation of new institutions is situated within the overall multifaceted state-building process, which results in transnational actors deploying side payments at both levels in order to achieve the primary goal of concluding an interstate agreement.

Since the Central Asian states were weakly institutionalized and poor, they were dependent on aid. Since they resembled quasi-states, they had a sense of entitlement to international compensation for ills suffered under colonial rule. The newly independent Central Asian states were not critical of aid, as their counterparts in Africa were, since they saw the Soviet Union—not the West, the World Bank, or the IMF—as their former colonial power. Wedel (1998, p. 17) similarly finds that the Central Europeans were eager to embrace Western aid because they could finally accept "the Western generosity that Stalin prevented them from accepting in 1947." The Central Europeans viewed programs such as the EU-sponsored Poland-Hungary Aid for Restructuring the Economy (PHARE) as a way to reenter Europe proper. Accordingly, the Central Asian successor states did not perceive internationally sponsored programs that promote rural development, structural adjustment, sustainable development, and basic human needs as an extension of the colonial power interfering in the domestic governance of these newly independent states. On the contrary, the Central Asian states also embraced the terminology of a Western development model as a means to shed the baggage of the Soviet model of development.[12]

Thus, IOs, bilateral aid organizations, and NGOs offer side payments at the international level in their expanded role in world politics as more than merely mediators and coordinators in the negotiation process. With the financial and material resources at their disposal, they possess the means to furnish side payments as selective incentives in order to shift the bargaining positions of the different actors with a stake in the new institutional agreement and to help to resolve collective-action problems.

Moreover, an investigation of a bargaining situation that deals with a physical resource demonstrates how technical and political aspects mesh. Many of the world's river basins are shared by developing countries. Since these states are often poor, they lack the technical capabilities and the financial resources to carry out water management schemes. In response, IOs provide side payments in the form of foreign capital to advance water development programs. Many NGOs play roles in transplanting know-how and ideas concerning water management practices from one setting to another. Their focus on the local level to spur community development projects is a means to contribute to societal transformation at a more macro level.

As was highlighted in chapter 2, international river systems are characterized by asymmetry of interests and capabilities. This is attributable to the physical structure of the basin, which influences the riparians' bargaining positions. For the purposes of the present discussion, what matters is that in cases of asymmetrical power relations between upstream and downstream riparians a potential exists for a third party in the role of an outside mediator to bring incentives and resources to the bargaining table in order to prevent a more powerful actor (the upstream riparian) from exploiting a less powerful one (the downstream riparian). In international politics those who gain by cooperation must devise incentives to make those who lose play the game, and this can then be done by means of side payments to those who stand to gain little by cooperation but whose cooperation is essential to joint management. In such cases, third parties still need to target specific domestic interests in both the upstream and downstream states that may want to alter the water-sharing arrangements.

After state breakup, the potential for this active role is even greater. With the disappearance of the center as the enforcer, an upper riparian no longer has incentives to uphold and recognize the mutual dependence of a shared water basin. From a purely physical and technological perspective, if the new upstream riparian (Tajikistan or Kyrgyzstan) had constructed a new permanent headworks, this would have further exacerbated the inequality among water users and reduced the recognition of mutual dependence among the riparians. What remains then is how to "equalize" the situation and re-create one of mutual interdependence after state breakup.[43] Indeed, another way in which IOs use the financial and material resources at their disposal to facilitate interstate cooperation is to sponsor various development projects among the states sharing the resource to either offset the physical asymmetry or to offer the new more powerful riparian an incentive not to exploit its new physical capabilities.

Besides serving as tangible incentives for states to negotiate new water-sharing institutions at the international level, side payments also play an essential role at the level of domestic politics. Transnational actors and government elites need to convince the multitude of domestic constituencies to abide by and fulfill these new institutional agreements. Consequently, some have suggested that the toughest bargaining is not among states but within them since an agreement may distribute benefits disproportionately across different groups within the state (Mayer 1992, p. 793). Thus, in order to persuade domestic actors such as bureaucrats and farmers not to sabotage any new agreements, the leadership may seek side payments for reasons tied specifically to the effects of state breakup and state formation. First, with the disappearance of the center, the

Central Asian governments lacked the financial and technical resources to provide socio-economic goods like environmental and social protection as well as patronage to republican and regional elites. Second, the sweeping transitions away from state socialism compounded the economic and social crises after state breakup, and the transfer of side payments helped the Central Asian leadership mitigate the harsh effects of the transition by enabling them to appease vital domestic constituencies during this period of domestic flux and uncertainty about the future.

Under conditions of transformation, new domestic governments prefer to tax the international community instead of their own populations, which again is wholly related to their status as weak or quasi-states. The use of external sources of rents or side payments enables the government leaders to meet their own domestic goals of consolidating independence, which includes breaking visible ties with the past system of rule. In response to this desire for foreign assistance, IOs, bilateral aid organizations, and NGOs are spending large sums of money and time trying to help post-communist states increase their domestic capacity to foster stability and to undertake hard choices by fully embarking on the transition away from state socialism. Transnational actors have been eager to provide assistance in many forms and to fill the void left by the disappearance of the center in Moscow. For example, in the environmental arena, they are offering capacity-building programs in order to strengthen the administrative and political abilities of these weakly institutionalized states (Haas, Keohane, and Levy 1993, p. 405). The Global Environmental Facility (GEF) is one example of a concerted effort by the UN Development Program, UNEP, and the World Bank to improve institutional capacity so that developing and post-communist countries will implement various global environmental accords.[44]

In order to support the manifold political, economic, and social transitions, IOs transfer financial and technical resources to these new governments to help them, for instance, introduce new legal institutions, tax systems, and banking institutions. Whereas in the past, IOs like the World Bank primarily assumed a more technical role, in the post-Cold War period other IOs, such as the European Bank for Reconstruction and Development, have taken on a political role to bring about the necessary institutions for sustaining democracy and markets.[45] These IOs are no longer neutral actors, but active actors with their own defined interests irrespective of the state actors involved in the bargaining situation over the formation of new institutions. Insofar as NGOs often lack abundant financial resources like the multilateral financial institutions, they will instead rely on their human capital to promote the growth of an active civil society as well as to provide services for basic human needs. At the same time, there is a growing trend for many financial institutions and bilateral aid organizations to "contract out" their projects to NGOs, thus giving the latter control over considerable financial resources.

In the aftermath of state breakup and the unsettling of boundaries, the side payments from IOs and NGOs serve as the impetus for institution building at both the international and the domestic level. In contrast with the conventional wisdom that perceives interstate cooperation as undermining state sovereignty, the active role of transnational actors in transitional states thus strengthens state sovereignty. Even though regional environmental cooperation was the overarching objective in the Aral basin, in order to reach that outcome transnational actors had to help constitute new states.

In summary: Transnational actors provide side payments as inducements for regional cooperation that in turn are used to compensate domestic constituencies. Thus, only an approach that stresses the interaction among weak domestic states, multiple third-party actors at the international level, and the introduction of side payments as the missing link can adequately account for rapid regional cooperation under conditions of transformation.

In order to ascertain whether my approach to two-level institution building is useful, in chapters 5 and 6 I explore the empirical implications of this approach for Central Asia. If it accurately depicts the situation of rapid regional institutionalization under conditions of transformation and over an international river basin, four things should be evident: There should be larger roles for IOs, bilateral aid organizations, and NGOs. Third-party actors should provide aid contingent on interstate cooperation. Domestic actors should welcome international intervention rather than resist it because it may violate their national sovereignty in the realm of policy making. Side payments should be channeled to the domestic level.

4

Cotton Monoculture as a System of Social Control

Historical Antecedents to the Aral Sea Crisis

Perhaps the most visible legacy of Soviet rule in Central Asia is the desiccation of the Aral Sea. The Aral Sea crisis embodies the failure of Soviet institutions to provide effective protection for the Central Asian environment and to ensure that the Central Asian populations received an adequate quantity and quality of fresh water to meet their basic human needs. Insofar as Soviet economic policies were directly responsible for the destruction of the Aral Sea ecosystem, the collapse of the Soviet Union along with the disintegration of its command-and-control economy provided a real opportunity to rectify past policies and restructure institutional mechanisms for water management. A potential for institutional change existed because state breakup destabilized Soviet decision-making structures that mediated relations between the center and the periphery. Precisely because of the structural-historical context,[1] however, institutional change is not a simple process in which institutions can be consciously designed. Rather, institutions tend to be quite conservative and resistant to change because fundamental change is a process with high fixed costs. Thus, it is often easier to maintain institutions than to adapt them or redesign them for new circumstances. When change does take place, it is usually path dependent (Stark 1992). Any modifications appear to be highly contingent on the previous institutional structure, even if it is inefficient (North 1990).

In order to put the post-Soviet bargaining situation over the introduction of new water institutions into context, this chapter lays out the multiple legacies of Soviet rule. If institutions can be designed only within certain structural-historical constraints, then we must know what those constraints are. This chapter does so by focusing on the multiple legacies of Soviet rule. Combined, these legacies were the initial constraints under which the Central Asian elites had to choose either to maintain previous institutions or to introduce new institutions. To begin, the Central Asian elites had to confront an environmental legacy that was tied to a grossly inefficient system of centralized water management. Because in the 1920s Soviet planners drew new political borders that did not correspond to the physical boundaries of the water system, the Soviet political legacy was another major impediment to devising new water-sharing institutions. Related to the political legacy is the legacy of Soviet nationality policy that grafted new "national" and "ethnic" identities onto existing societal cleavages. Finally, and most important, Soviet planners introduced a system of economic regional specialization in which cotton production dominated the Central Asian economy. The Soviet system of cotton monoculture, in turn, begat a system of social control based on reciprocal patronage relationships among Moscow, the regional and local elites in the republics, and the general population. The legacy of cotton monoculture as a system of social, political, and economic control has prevented the Central Asian elites from choosing the most effective institutions for rectifying the environmental situation in the Aral basin in the post-Soviet period.

This chapter explicitly compares the Soviet period with the Czarist period in order to shed light on the changes in water institutions over time. Policy choices made during those periods are the historical antecedents for post-Soviet bargaining over the formulation of new water-sharing institutions in the Aral basin.

The Russian Conquest and the Maintenance of Traditional Institutions

Cotton and Political Incorporation into the Russian Empire

On June 15, 1865, Russian troops belonging to the Fourth Orenburg Line Battalion arrived in Toshkent, marking the beginning of Russian rule in Central Asia (then referred to as western Turkestan).[2] One of the main objectives of Russian colonial expansion was to encourage cotton production specifically in the oases among the sedentary population in order to bolster the growing textile industry in Central Russia. With the outbreak of the American Civil War (1861), Russia found itself cut off from its main market for cotton, as did the rest of Europe. In direct response, fifteen Muscovite merchants asked the minister of finance for assistance to help them to procure the raw commodity from Central Asia. Cotton manufacturing then evolved into one of the leading branches of the Russian textile industry, particularly in the province of Moscow (Carrère d'Encausse 1994, p. 131).[3]

The Russian conquest of Central Asia had numerous implications for the region's geopolitics, its economy, and its socio-cultural conditions. The Russian authorities divided the inhabitants of Central Asia into two overarching geographical areas, which were seen as coterminous with two distinct cultural traditions and economic regions: the steppes and the oases.[4] Central Asia was politically delineated into the Guberniia (governorate-general) of Turkestan on July 11, 1867, and the following year the Guberniia of the Steppe was organized.[5] This distinction between the steppes and the oases was carried over into the Soviet period, resulting in the Soviet authorities' treatment of the region as two distinct regions: Kazakhstan and Central Asia. The latter referred to Uzbekistan, Tajikistan, Turkmenistan, and Kyrgyzstan. This demarcation between the steppes in the north and the oases in the south affected the governance of Central Asia and the Soviets' later encouragement of different sectors of regional economic specialization.

Traditional Methods of Water Utilization in Pre-Soviet Central Asia

When Czarist troops entered Turkestan in the mid 1800s, they encountered decentralized patterns of governance in multiple domains at the village level. Rather than seek to dismantle local institutional structures, they left many intact not only for administrative purposes but also as a mechanism of social control. For example, they turned to the traditional village elder (aqsaqal) to regulate everyday interactions, especially in the agricultural sector that was the basis for the local economy. Specifically, Czarist administrators retained local irrigation practices to foster settled agriculture and provide for the planting of cotton in the oases. These traditional irrigation practices in pre-Soviet Turkestan later became the focal point of Soviet planners' actions to re-engineer the Central Asian water system.

Irrigation is a fundamental component of the expansion and maintenance of agriculture throughout the world. Irrigation has always been essential for settled agriculture to take hold in semi-arid and arid environments. Even under conditions where water is scarce and highly seasonal, early civilizations were able to devise irrigation systems that could exploit underground water tables. In Iran and other parts of the Middle East and Northern Africa, underground water canals (qanats) supported agriculture (Beaumont, Bonine, and McLachlan 1989). Such local irrigation systems enabled numerous cities, towns, and villages to flourish for thousands of years in seemingly unfavorable conditions. This form of irrigation was also practiced in Turkmenistan (where it was called a kariz) near the base of the alluvial fans along the foothills of the Kopet Dag mountain chain.

Similar to other hydraulic civilizations of the Middle East, the oasis regions of Central Asia have a long history of indigenous agriculture and irrigation use, dating back several thousand years. Akin to other Muslim societies, the traditional Central Asian system of water rights was based, in theory, on Islamic practices and customs according to which water belonged to the whole community (i.e., the state or the public domain) and could not be bought and sold (Caponera 1992, pp. 68–69).[6] In Central Asia, however, most irrigation took the form of an above-ground water system of main canals and secondary offshoot or lateral canals. When the Russians arrived in Turkestan, they discovered an extensive network of canals in the oases of the Fergana Valley, along the Zarafshon, lower Amu Darya, the Murghab, and the Tejen (Matley 1994). In Khiva alone there were six major canals, ranging in length from 72 to 96 kilometers (Wheeler 1966, p. 32).

In contrast to the nomadic peoples of the steppes, who subsisted through hunting and animal husbandry, villages in the oases grew as a result of the development of local and stable irrigation systems that met basic social needs; in most cases, the technology was appropriate to the environment. At the same time, a decentralized system of irrigation kept populations distinct and separated. Populations of neighboring villages (qishlaqs) hardly knew one another, especially if these villages were located on different main canals (Khazanov 1992, p. 78). Even in situations where different oases were located on a similar river, groups rarely intermingled, since they were usually far enough physically removed from one another (ibid., p. 82). Overall, this meant that cultivators seldom had to compete for access to water resources.

Traditional methods of farming were able to uphold the fertility of the soils by preventing water logging and exhaustion of the soils. Farmers surrounded their fields with low earthen walls and mulberry trees to facilitate drainage runoff.[7] Water norms were tied to the rotation of crops (wheat, barley, millet, alfalfa).[8] Alfalfa, in particular, was essential for maintaining the richness of the soil, owing to its nitrogen-enriching properties; cultivators traditionally then rotated alfalfa with cotton and, in turn, would use it as fodder for livestock. The oases were especially renowned for their melons, apricots, peaches, figs, cherries, pomegranates, apples, almonds, and other fruits and nuts. Another mainstay of the local economy was the mulberry tree, whose leaves were used to raise silkworms.

Similar to other political and social institutions that were organized around village life, so was the traditional method of water management. For agriculturists dependent on irrigation for their livelihood, the maintenance of a canal and the regulation of the water flow are extremely critical.[9] This, however, requires collective action among the members of the community. An extensive system of rules and practices evolved to govern the use of the canals and allocation of the irrigation water throughout the oases of Turkestan.[10] In practice, at the level of the volost (a gathering of several villages or qishlaqs), a local water master (mirab) was appointed to supervise the construction and maintenance of a major canal and its system of distribution canals and irrigation ditches.[11] He ensured that each landholder received the flow of water for the number of days to which he was entitled. In most cases, a village elder (ariq aqsaqal) oversaw the mirab.[12] Overall, Russian intervention in these local institutions was minimal; rather, the Russians permitted and moreover relied on the former local judicial and political institutions of the village to govern everyday life. Thus, the village leaders (aqsaqals) were able to function in their traditional role as locally elected officials.

The upkeep of a canal requires a great deal of labor, especially to remove the silt and other debris that accumulates. From Khazanov's research, it is evident that the demand for collective action did not result in centralized institutional arrangements "in spite of the fact that the maintenance of irrigation networks and the building of new canals usually demanded a certain degree of governmental management and participation" (Khazanov 1992, p. 83). In most cases, canals and ditches (usually furrows) that irrigated individually owned fields were the responsibility of the farmer. In contrast, the cleaning of a main canal became a communal endeavor. All families using a canal were expected to contribute labor to cleaning and maintenance. Here again the mirab played an important role, seeing to it that each family sent sufficient labor and supplies on the designated day.

Even though the growing of cotton expanded in the mid 1800s, farmers largely preserved their traditional patterns of rotation and methods of water utilization until the Soviet period as a result of the Russian colonial policy of non-interference or what can otherwise be referred to as indirect rule. Yet with Russia's burgeoning desire to free itself from dependence on foreign supplies, cotton cultivation began to replace many of the traditional crops in the Fergana and Zarafshon Valleys. By 1909, 25 percent of the irrigated area in the Zarafshon Valley was devoted to the planting of cotton (Matley 1994, p. 275). This pattern accelerated during the Soviet period, having a tremendous and deleterious effect on the political and social aspects of water utilization and on the physical environment in the Aral basin.

The Bolshevik Revolution and Institutional Change

Political Incorporation into the Soviet Union

For the territory of Turkestan, the Bolshevik Revolution brought widespread institutional change, altering the political, economic, social, and physical constitution of the region. In a study of irrigation institutions in France, Rosenthal (1992) argues that reforms carried out during the French Revolution improved both the efficacy of water control and the overall structure of property rights in agriculture by destroying the old system of privilege. On the contrary, the Bolshevik Revolution and subsequent changes in property rights over land and water coupled with centralized planning created one of the most inefficient systems of water management and one of the most inefficient agrarian-based economies in the world.

Both in contrast to and in response to Czarist policies, Bolshevik policies in Turkestan had a profound and lingering impact not only on the structure of land and water rights but also on political and social cleavages in the region. These legacies of Soviet rule undergird post-Soviet Central Asian politics, since the Central Asian successor states had independence thrust upon them through a process of involuntary state breakup. They were not formed through war or through political struggle. Not surprisingly, many of the post-Soviet bargaining positions concerning water sharing are conditioned by early Bolshevik policies that constructed a situation of mutual dependence by the mere way political borders were drawn and through the means in which the physical system was exploited for the promotion of a monocrop economy.

The conflict between political and economic boundaries is a direct result of the way in which the Bolsheviks divided up Central Asia. At the time, they had two main priorities: nation building and economic development. However, the economic priorities ranked below the national ones, and these priorities often conflicted. In fact, because the national goals took precedence, irrigation systems had to cross administrative-territorial boundaries.

The shifting of territorial boundaries in the early Soviet period shapes the bargaining situation over the formation of water-sharing institutions in the post-Soviet period, since these Soviet crafted republican borders became the inherited political borders of new sovereign states. The Bolshevik strategy for the incorporation of Turkestan into the Soviet Union was based on the replacement of a united Turkestan with smaller national republics. The Bolshevik authorities abrogated previous Czarist administrative territorial delineations and rearranged the borders several times before settling for the borders of five Central Asian republics, namely Uzbekistan, Kyrgyzstan, Kazakhstan, Turkmenistan, and Tajikistan. Even though these republican borders did not remotely correspond to any other previous political boundaries, the designers of the new borders did not draw them arbitrarily but rather relied on the work of ethnographers who identified various ethnic and linguistic cleavages among the populations (Slezkine 1994; Jones Luong 1997, chapter 3). In short, the division of Turkestan into five Central Asian republics was part of a Soviet strategy to create new national identities tied to territorial units that could then both be used as means for rule and for incorporation into the larger Soviet Union.[13]

The territorial reorganization of Turkestan officially began in October 1924 when the Central Executive Committee of the Soviet Union established two socialist republics: Uzbekistan and Turkmenistan. Uzbekistan encompassed the central part of old Bukhoro, the southern part of old Khiva, and the regions of Samarqand, Fergana, Amu Darya, and Syr Darya. Turkmenistan comprised the Turkmen regions of western Bukhoro, Khwarazm, and the former Trans-Caspian region. At this time, Tajikistan and Kazakhstan were designated autonomous republics. Tajikistan covered the mountainous regions of former eastern Bukhoro (includes present day Gorno-Badakhshan), where an essentially Shiite Iranian-speaking population lived, but did not include the predominately Tajik cities of Bukhoro and Samarqand; on December 5, 1929, it was elevated to rank of a federated socialist republic. The Kyrgyz Autonomous Oblast became the Kyrgyz Autonomous Republic on May 25, 1926. On December 5, 1936, the Kyrgyz Republic and Kazakhstan obtained the status of federal socialist republic. Also in 1936, the Karakalpak Autonomous Soviet Socialist Republic was transferred to the Uzbek SSR from Kazakhstan.

The Bolshevik authorities used the new national delimitations to try to supersede previous social identities that did not coincide with "proper" ethnic and/or linguistic markers. For example, the Sarts (a name given to sedentary people in the oases) were now referred to either as Tajiks or Uzbeks. Although these federal and autonomous socialist republics were each named for a titular population, political borders did not necessarily fully concur with the titular population; in most cases, these republics were ethnically diverse, and they became further intermingled throughout the Soviet period. Besides the various indigenous Central Asian populations, other ethnic groups were deported to Central Asia in the ensuing decades. For example, during World War II and shortly thereafter Stalin ordered mass movements of Koreans, Crimean Tatars, Meskhetian Turks, and even Volga Germans to Central Asia. In short, Soviet nationality policies created republics characterized by intermingled ethnic groups. (See table 4.1.)

Table 4.1
Ethnic breakdown of Central Asian republics, 1989 (thousands). Source: Pomfret 1995, p. 5.

	Kazakhstan	Kyrgyzstan	Tajikistan	Turkmenistan	Uzbekistan
Kazakhs	6535	37	11	88	808
Kyrgyz	14	2230	64	1	175
Tajiks	25	34	3172	3	934
Turkmen	4	1	20	2537	122
Uzbeks	322	550	1198	317	14,142
Russians	6228	917	388	334	1653
Ukrainians	896	108	41	36	153
Byelorussians	183	9	7	9	29
Germans	958	101	33	4	40
Tatars	328	70	72	9	657
Karakalpaks	—	—	—	—	412
Koreans	103	18	13	—	183
Uigurs	185	37	—	—	36
Total	16,563	4290	5109	3534	19,905

The initial shifting of borders and movements of peoples directly affected the bargaining over the formation of water-sharing institutions in the post-Soviet period by creating a situation in which multiple heterogeneous users became dependent on a similar water resource. The clearest example is the case of the Fergana Valley. The Soviet authorities carved up the Fergana Valley among three republics: Kyrgyzstan, Tajikistan, and Uzbekistan. Within a territory 300 km in length and 20–70 km in width, the political borders weave in and out, leaving potentially irredentist populations outside their titular republics.

More important for the question of water sharing is that, even where the Soviet government attempted to divide Central Asia politically and nationally, it paradoxically sought to physically and economically integrate the Aral basin by encouraging regional sectoral specialization. Most of the region of Soviet Central Asia (approximately the Guberniia of Turkestan) fit within the boundaries of the Aral basin, and this is where cotton cultivation was zealously promoted. In contrast, most of the territory of Kazakhstan (formerly the Guberniia of the Steppe) did not accord with the geographic boundaries of the Aral basin; only the two southern oblasts in Kazakhstan (Shymkent and Qyzlorda) are situated within the physical boundaries of the basin. The soil and climate of the steppes was not conducive to the growing of cotton, and instead most of Kazakhstan's agriculture was geared toward animal husbandry and grain production in the north. Such geographic markers (a holdover from Russian rule) reinforced the Soviet Union's designation of Kazakhstan in 1961 as a separate economic district (Lipovsky 1995, p. 539).

The Introduction of Cotton Monoculture and the Expansion of Irrigation

In an article on the politics of irrigation in Czarist Russia, Joffe (1995, p. 367) notes that the Soviet authorities were later able to carry out many of the recommendations that the Russian cotton industrialists had hoped to execute during the early 1900s. At that time, they had wanted to expand the irrigation system in Turkestan so that more land could be brought into cultivation; their efforts did not bear fruit because the Czarist government could not guarantee basic property rights and the rule of law. Instead, Soviet planners undertook the economic integration of Turkestan through centralized means rather than with private capital. In spite of a deep-rooted tradition of irrigated farming in the oases, Soviet planners set out to intentionally transform local methods of water use on a colossal scale to meet the imperatives of a Soviet economic system predicated on regional specialization.

Central Asia, with its hot climate, was to provide the bulk of the Soviet Union's raw cotton, and irrigation was the means by which it was to achieve self-sufficiency in cotton production. In May 1918, Lenin issued a decree "about the organization of irrigation work in Turkestan" which then provided the basis for the introduction of large-scale irrigation projects to ensure the independence of cotton imports (Lenin and Stalin 1940, pp. 54–59).[14] Before the Soviets could expand cotton cultivation, however, there were several obstacles to be overcome. First, the local economy was collapsing as a result of the civil war; second, Russia had ceased its grain shipments to Turkestan during the civil war, causing many peasants to abandon growing cotton for grain and other traditional crops in order to obviate the effects of widespread famine. In order to convince the peasants to plant cotton for processing in Russia's textile industry, the Soviet authorities passed a series of resolutions aimed at the restoration of the Central Asian irrigation system which had fallen into a state of disrepair. For example, in 1923 the Soviet authorities established reclamation cooperatives to assist local farmers by offering credits to help them rebuild their own irrigation systems (Matley 1994, p. 287).

Fundamental institutional change requires a significant departure from past practices, and clearly the Soviet authorities sought to alter and replace traditional land and water practices in the agricultural sector. Moreover, they endeavored to transform Central Asian society through the restructuring of the traditional economies. By taking away the authority of the mirab and centralizing decision making about water allocations and plantings, the Communist Party aimed to firmly establish a presence among the indigenous populations. Clearly these actions were intended to compel Central Asian society, especially in the oases, to meet the growing demand from the center for raw cotton exports. Some suggest that these reforms were also used to undermine the role of Islam and customary practices in the villages as preparation for the collectivization campaign and other programs intended to weave the fabric of a new enlightened "socialist" society (Wheeler 1966, p. 73).

From a purely instrumental perspective, the restoration and subsequent enlargement of the irrigation system served two purposes for the Bolsheviks. First, it helped them to meet the economic challenge associated with increasing cotton productivity. Second, it provided the means to appease the peasants in order to bring them under more centralized control and to accelerate the process of collectivization. In Uzbekistan, for example, where early attempts at collectivization had largely failed, the Soviet authorities initiated land and water reforms between 1925 and 1929, making newly reclaimed land available to those willing to accept collectivization.[15] In 1924 there were only 62 kolkhozes in all of Uzbekistan; by 1927 there were 832.[16]

Changes in land tenure patterns, in property rights, and (particularly in Central Asia) in water-use practices are critical for understanding state-society relations in agrarian societies because these institutional changes become the new rules of the game for society (Migdal 1988, p. 57). Whenever there is a turnover in leadership, the new elites will target these institutions first, since they define the basic strategies for survival in societies where agriculture is the primary mode of sustenance. Soviet policies of collectivization and centralization sought to undermine traditional social power relations predicated on patron-client relations within the villages or kinship ties in the steppes and replace them with new superstructures associated with the kolkhoz (collective farm) and the sovkhoz (state farm) at the micro level and with administrative-territorial districts at the macro level associated with the rayon and the oblast. Attempts at collectivization were also used to settle the nomadic peoples (Kazakhs, Kyrgyz, and Turkmens) by disrupting traditional patterns of seasonal migration between the aul (summer settlements) and the uru (winter settlements).[17]

Yet Soviet planners unwittingly reinforced many pre-existing patronage relations in Central Asia within these new forms of social organization at the micro level by often grafting these new superstructures onto pre-existing villages or onto groups with similar kinship structures (Winner 1963b). In order to fortify many of these new collective and state farms, the Soviet authorities allowed them to retain the summer herding of animals along the lines of a "brigade," for example. By simply overlooking the "persistence" of various elements of traditional patterns of social organization, this helped consolidate the settling of the nomadic peoples after the initial attempts at forced collectivization had encountered much resistance.

Along with the collectivization campaign in the 1930s, the Soviet authorities needed to expand irrigation to support many of these new farms. In accordance with the Soviet emphasis on "bigness" and the "monumental," the engineers embarked on the construction of several of the major canals in the Fergana Valley that would link together the various oases. The largest of these canals, the Great Fergana Canal, was built in the summer of 1939 by "voluntary" or "unpaid labor" consisting of 160,000 bodies (primarily Uzbek) from 2140 kolkhozes representing 18 rayons in the Fergana Valley; these manual laborers who were accompanied by the supervising engineers and technicians (primarily Russian) were able to complete the digging of this 249-km canal in a record 45 days (Matley 1994, pp. 294–295).[18] Afterward the canal was extended into Tajikistan. The Soviet authorities mobilized the population into work brigades, rallying them around the slogan that they were participating in the construction of socialism; yet these practices were often no different than those employed under the khanates to garner unpaid labor.

The Great Fergana Canal serves as the basis for the whole irrigation system within the Fergana Valley; it is fed by the Naryn and Kara Darya rivers and runs the length of the valley from northeast to southwest. Other smaller, but important major canals built during this period include the North (133 km) and the South Fergana (93 km) canals and the Savai Canal (53 km) in Andijon. In Tajikistan, the first stage of the main canal in the Vakhsh Valley was completed in 1934.

Although World War II interrupted many of these projects, by the 1950s capital investments once again swelled for the expansion of the irrigation system and the simultaneous reclamation of new lands for agricultural production of cotton. During this era, Soviet planners promoted large-scale engineering projects that further glorified the radical and grandiose "transformation" and "modernization" of the rural and traditional sectors of Central Asian society. This mirrored Stalin's great plan to transform and subjugate nature for economic gain irrespective of ecological and human costs (Rostankowski 1982; Weiner 1988). In practice, this led to the development of new marginal lands even when the economics of cost-benefits would have suggested otherwise. Before the 1950s, irrigated agriculture was concentrated, for example, in Uzbekistan in Bukhoro, Fergana, and Khorazm; now Soviet planners turned their attention toward increasing agricultural output in regions such as Qarshi, Jizzakh, and the Golodnaya (literally, hungry) Steppe. At the same time, the reorganization of traditional irrigation networks in conjunction with the extension of new ones facilitated the process of linking together the collective and state farms in a situation of mutual dependence.

Among the first major endeavors of this period were the reclamation projects in the Golodnaya Steppe, a region located between Toshkent and Samarqand.[19] Many previous attempts to turn this barren desert into fertile fields had failed during the Czarist period, with the notable exception of the completion of the Romanov Canal (later called the Kirov Canal) in 1913, which became the main magisterial canal in a network to supply the Golodnaya Steppe with water.[20] Canals such as the Southern Golodnaya Steppe canal were built to further augment the amount of cultivated acreage in the 1950s and the 1960s. Once again, after the delivery of water from the Syr Darya to the Golodnaya Steppe, the Soviet authorities needed to encourage the migration to and the settlement of the steppe to farm these lands. Many disparate and non-indigenous populations such as the Koreans were either forcibly relocated to the region or lured there because of certain benefits and privileges offered by the government.[21] Unlike other kolkhozes that were built as extensions onto older villages, these new farms were constructed according to a standard plan that included specific details such as for housing and municipal services. Although the reclamation of the Golodnaya Steppe was clearly geared toward increasing the overall level of cotton production in the Soviet Union, the settlement of new collective and state farms in regions like the Golodnaya Steppe served as a base for the expansion of patronage networks through the growing of cotton, regardless of the ethnic composition of the farms.

Another prototype of these large-scale projects that emphasized the development of new lands without taking into account their most efficient use was the construction of the Kara Kum Canal in Turkmenistan. The first stage of the canal was completed in 1959 when it reached the oasis of Mary. As the longest canal in the world, extending for more than 1300 km, it initially diverts water from the Amu Darya at Kerki near the Afghan border and then transports it across the desert though Ashgabat, the capital of Turkmenistan. Whereas originally it was intended to reach the Caspian, by the 1980s the canal was extended as far as Kazandzhik, whereupon shortly afterward it stops dead in the midst of the desert. The protracted elongation of the canal allowed Turkmenistan to develop more than 850,000 hectares of additional irrigated land between 1960 and 1990.[22] It should be noted, however, that all water diverted to this canal is lost to the Aral Sea.

Unlike the pre-Soviet irrigation systems that were based on local institutional arrangements and tied to a particular water source, Soviet policies linked different users across different drainage basins (Gleason 1991). By launching these massive development projects, Soviet planners engineered a highly integrated irrigation system with intra-basin and inter-basin water transfers. Soviet planners managed to create a situation of physical interdependence for the entire region. This fostered a situation of competition among agricultural, municipal, and industrial users over the quality and quantity of the river flow in different regions.

Consider the Syr Darya basin, for example. Here, upstream reservoirs and dams in Kyrgyzstan (including the Toktogul hydraulic complex) are connected to downstream irrigation systems for agriculture at the farm level in both Uzbekistan and Kazakhstan. These reservoirs, built primarily to help control and extend irrigation along the whole length of the Syr Darya to the Aral Sea, are also capable of generating hydroelectricity. After the water leaves the upper watershed, the first user in a chain of agricultural users dependent on the Syr Darya water system is the Fergana Valley, where most of the cotton growing and other agriculture takes place. From the Fergana Valley the Syr Darya continues to flow downstream, where it is siphoned off into the various canals for application in the Golodnaya Steppe. The water leaves Uzbekistan at the Chardara reservoir and flows through Kazakhstan to the Aral Sea; however, this water is of a much poorer quality since a substantial amount consists of polluted drainage water that has been returned to the river. Contaminated with fertilizers and other chemicals, this water from the Syr Darya is then utilized for both agriculture and drinking along the Syr Darya in Kazakhstan.

A similar situation exists in the Amu Darya basin. In the upper reaches of the Vakhsh River, the Nurek hydroelectric complex is the main regulator of the flow of the upper tributaries of the Amu Darya. Water released from the Nurek hydroelectric complex irrigates land first in Tajikistan. The next main water users are other agricultural users in Turkmenistan after the Amu Darya is diverted from the Kara Kum canal. Downstream users in Khorazm and Karakalpakstan in Uzbekistan then apportion the remaining water left in the Amu Darya before it reaches the Aral Sea.

By the mid 1980s, owing to the expansion of irrigation, there were approximately 7.2 million hectares of irrigated land in Central Asia, more than half of which was located within Uzbekistan; in contrast, in 1950 there had been only 2.9 million hectares, increasing dramatically to about 5 million hectares in 1960 (Glantz, Rubinstein, and Zonn 1994, pp. 167–168). Irrigation accounts for about 85 percent of all water withdrawals in the basin.[23] Hence, most of the arable land in Central Asia is irrigated. (See table 4.2.) In Uzbekistan, the importance of irrigation for achieving

Table 4.2
Data on arable land and irrigation in Central Asian republics (whole-country estimates) in the early 1990s. Source: "Irrigated Crop Production Systems, Volume IV," TACIS, p. 3.

	Kazakhstan	Kyrgyzstan	Tajikistan	Turkmenistan	Uzbekistan
Arable area (millions of hectares)	39.6	1.40	0.81	1.27	4.50
Percentage irrigated	6.1	76.4	86.3	100	93.3

cotton independence is immediately discernible: Only 10 percent of the land is cultivated, but 95 percent of that land is irrigated (World Bank 1993b, p. 115). By the early 1980s, the Soviet Union was the second largest cotton producer in the world (behind China), accounting for close to 20 percent of the world's production (Rumer 1989, p. 62). To achieve economic autarky, Soviet planners thus allocated most resources (including water) in Central Asia to develop and support a monocrop economy, not taking into account the social costs in terms of the health of the population and quality of the environment.

The transformation of cotton into a monoculture crop is one of the main reasons for the environmental crisis in Central Asia (Rumer 1989, p. 70). Cotton cultivation evolved into the dominant economic activity in Turkmenistan, Tajikistan, and Uzbekistan where it was grown predominantly on the collective and state farms. In Uzbekistan alone, the cotton sector produced more than 65 percent of the republic's gross output, consumed 60 percent of all resources, and employed approximately 40 percent of the labor force by the mid 1980s; the republic accounted for approximately two-thirds of the cotton produced in the Soviet Union (ibid., p. 62). Another source finds that by the beginning of the 1990s the annual cotton harvest in Central Asia had reached 8 million tons (Lipovsky 1995, p. 534).[24] Even though Central Asia's role in the overall Soviet economy was to produce the raw material, any processing beyond the ginning stage took place elsewhere, usually in Central Russia in cities such as Ivanovo.[25]

As a consequence of regional specialization, other segments of the economy were sacrificed to sustain high cotton yields. Even though the expansion of irrigated land increased the size of the cotton harvest, it also created a downward spiral wherein to produce more cotton to meet the growing demands from the center, more land constantly needed to be brought into cultivation. Yet the newer lands were usually the least suitable for cotton growing. Soviet planners halted traditional rotations, and instead relied on the application of endless amounts of fertilizers along with various pesticides and defoliants (such as the later banned butifos) to aid the growing and harvesting periods. Since the entire system of incentives was directed toward achieving higher and often unattainable targets, the quality of the cotton in Central Asia declined by the 1980s as soils became exhausted and salinized. The most conspicuous effect of the expansion of irrigation and monoculture in Central Asia is the desiccation of the Aral Sea. As irrigation became the dominant user of water in Central Asia, only a trickle of water reached the Aral Sea during the 1980s.

Centralized or Coerced Institutions for Water Management in Soviet Central Asia

A system of irrigated agriculture is "a landscape to which is added physical structures that impound, divert, channel, or otherwise move water from a source to some desired location" (Coward 1980, p. 15). According to Rakhimov (1990, p. 7), in 1987 there were 967 irrigation systems, 915 hydraulic structures, and 260 dam water-intake systems in Central Asia. In Uzbekistan alone, there were about 170,000 km of canals that irrigated 4.2 million hectares of land (IMF 1992, p. 1).[26] Most of the irrigation is of the furrow or flood type, where water is drawn from larger supply channels or magisterial canals. The canal systems in Central Asia use gravity as the primary force in moving water down slopes and pumping stations to raise the water for transport. Unlike the system of small-scale irrigated agriculture in the pre-Soviet period where it was feasible for local water authorities to supervise water withdrawals and the maintenance and cleaning of the canals, this large-scale and highly integrated system required careful coordination among multiple users across various sectors with competing interests. Hence, coordination was imperative to regulate the flow among them, to provide for the upkeep of the hydraulic structures, and, most important, to mitigate potential conflicts over water quality and quantity.

Previously, a central hegemon existed in Moscow, solving the collective-action problem over the sharing of this large-scale common-pool resource by imposing a water regime on the Central Asian republics. In a normative sense, Soviet centralized institutions supplied the mechanism for final decision making while at the same time they provided stability by resolving much of the internal conflict among the republics. Even when disputes transpired at the all-Union level (especially over the last 10 years of the Soviet Union) concerning the decreasing water flow in the rivers, the Soviet leadership or the various economic ministries involved in water policy (i.e., the Ministry of Agriculture or the Ministry of Energy) still had the final say on water management decisions.[27]

If institutional arrangements set the rules that coordinate actions (i.e., solutions to collective-action problems) among various water users, what in particular was it about the Soviet institutional structure that created a water management crisis, culminating in the desiccation of the Aral Sea? Moreover, how could this imposed institutional structure simultaneously foster cooperation among the competing users and relative stability among the population while at the same time being highly exploitative?

The main organization for managing water distribution in the Soviet Union was the Ministry of Land Reclamation and Water Resources (Minvodkhoz), located in Moscow.[28] Similar to Islamic water practices in which water belongs to the whole community, Article 6 of the Soviet Constitution affirmed that "waters, as the land, the subsoil and the forests, are the property of the Socialist Soviet State, that is, a matter of all people" (Caponera 1992, p. 85). The Bolshevik Revolution transferred ownership from the khanates to the Soviet state. Minvodkhoz then acted on behalf of the state as its steward.[29] Two properties defined the system of Soviet water law. First, water remained the exclusive property of the state. Second, and integrally related to the notion of state property, water was non-alienable (Davis 1971, p. 65; Kolbasev 1971). Only the right to use water could be assigned.

The Soviet system for water management in Central Asia was indicative of the general problems associated with a command-and-control economy for the allocation of scarce resources; it was highly centralized, vertically integrated, and extremely bureaucratic. In theory, all decisions related to the various aspects of the water sector emanated from Moscow starting with the State Planning Committee (Gosplan) and the various affected government ministries that established production targets and then allocated inputs based on the planning and implementation of specific programs to promote "economic development." As part of this system, Minvodkhoz determined the timetables and the amount of water allocated for irrigation between the upstream and downstream republics after reviewing the forecasts that the Main Hydrometeorology Service (Glavgidromet) made each year (Buck, Gleason, and Jofuku 1993). In contrast to a decentralized economic system in which allocations are a function of supply and demand and the generation of profit, here market forces did not sway allocative decisions and/or affect a ministry's survival; on the contrary, few ministries or agencies disappeared as a result of poor performance. In fact, a false market existed in which both land and water had no value; in the absence of markets, water was highly subsidized and consequently appropriated freely and generously.

The management of water resources shared by two or more republics was further tied to various territorial schemes for "the complex use and protection of water." During the 1980s, the center sought to shift some of the decision-making authority to the republican ministries, allowing them to coordinate and formulate the plans for shared resources together and then submit them to the all-Union Minvodkhoz for final approval. For example, the scheme for the use of the water resources of the Syr Darya basin was devised in 1983 and then further updated in 1987.[30] In the late 1980s, to augment these schemes, Soviet planners created two basin-wide agencies for water allocation (Bassenovoe vodnoe ob'edinenie, or BVOs), one for the Amu Darya basin and one for the Syr Darya basin. Subordinate to Minvodkhoz, they were to have an executive function—to manage a cascade of reservoirs, water-withdrawal facilities, and pumping stations. Soviet planners intended for the BVOs to control and monitor the flow of the rivers and allocate water to the different irrigation canals and direct users in the region. The regional center for the distribution of water from the Amu Darya was located in Urgench (Khorazm), Uzbekistan; that for the Syr Darya was in Toshkent, Uzbekistan.

In calculating the water allocations, Soviet planners rarely consulted the local water administrations or the farmers. On the contrary, decisions made at the highest levels were supposed to slowly filter down through the system to the oblast level, then to the rayon level, ultimately affecting distribution patterns at the farm level.[31] This was made possible by the physical construction of a centralized irrigation system that provided water to the collective and state farms though a chain of canals and hydrotechnical structures (Kolbasev 1971, p. 125). The final execution of these decisions rested with the irrigation administrations at the local level (raionnye upravleniia orositelnykh sistemi) that were responsible for distributing the irrigation water among the farms.[32] Minvodkhoz's jurisdiction ended at the farm level once the gates were opened and water released to the farms. The particular limits for the individual farms were then worked out at the local level between the kolkhoz or sovkhoz and the local irrigation administrations.[33] Even though the farm would submit its overall plan to the local Soviet (rayispolkom) for approval, the individual farmers themselves were completely disassociated from decision making, as they were only one aspect of the larger system devoted to maximizing agricultural production, namely cotton. Since Minvodkhoz's jurisdiction did not extend to the farm level, this created a gap between what happens at the last official point and where the water enters the farm or the fields. Yet all maintenance and repairs of the intra-farm network rested with the farms. As a result, Minvodkhoz and its local administrations possessed limited means to monitor the efficiency of use at the farm level since the farms, first and foremost, did not have to pay for the water received (Gustafson 1981, p. 129). Water gauges were absent and water was basically delivered via above-ground, open, and unlined canals which made it difficult to regulate water use for each individual.[34] The only somewhat reliable figures that exist on water usage in Central Asia pertain to the delivery of water to the administrative district areas and in some cases to the farm gate. Compounding the inadequacy of oversight, farmers furthermore lacked individual incentives to use water efficiently since they were only rewarded for meeting production quotas rather than maximizing efficiency and conserving water.

Another critical reason for the paucity of water conservation in Central Asia was that Minvodkhoz's primary interests were to build irrigation and drainage systems for the reclamation of new lands. Indeed, Minvodkhoz was not just responsible for water allocation decisions, but through its many subsidiaries, it possessed the means to undertake the above mentioned huge civil works projects. Tied to Minvodkhoz were numerous construction trusts and engineering design institutes, scientific research institutes such as the Central Asian Scientific Research Institute for Irrigation (SANIIRI), and training institutes like the Toshkent Institute for Water Engineers and the Mechanization of Agriculture. For example, the Ferganavodstroi Trust of the Uzbek Minvodkhoz built the early irrigation networks in the Fergana Valley.

Since Soviet planners emphasized the rapid completion of plans, this resulted in the poor construction of many of the installations. Moreover, many of the projects were never finished and remained half-built. Gustafson (1981, p. 125) observed that "in reclamation as elsewhere in the Soviet economy, the incentive system rewards gross output—so many of kilometers of canals, so many tons of earth moved—rather than the number of completed projects turned over to state and collective farms, let alone their contribution to improved yields." Numerous examples of unfinished projects dot the landscape in Central Asia. The Kara Kum canal, originally intended to reach the Caspian Sea, stops in the midst of the desert; a series of reservoirs (Kambarata) remain unfinished in the upper reaches of the Naryn River in Kyrgyzstan; the Rogen hydroelectric complex in the upper reaches of the Vakhsh River in Tajikistan was still under construction as the Soviet Union collapsed.

Poor and hasty construction underlay much of the water inefficiency in Central Asia. Many of the canals were unlined, which led directly to the loss of large quantities of water through seepage and evaporation. According to one source, fewer than 10 percent of the canals in Uzbekistan were lined with polymer or concrete material as of 1987, and in Turkmenistan the percentage was thought to be even lower (for both intra and inter-kolkhoz canals) (Chalidze 1992, p. 31). And many of the older irrigation systems lacked proper drainage networks, which contributed to water logging and salinization of the soil. At independence, about 2.1 million hectares in Uzbekistan were considered to be seriously damaged by salinity, and most of these were in the lower reaches of the basin (World Bank 1993b, p. 128). Policies favoring new projects over reconstruction (e.g., lining many of the earlier canals) allowed for such inefficiencies.

Finally, the disincentive to use water prudently is integrally related to the overlapping and often conflicting jurisdictions existing within the mandate of the same ministry and across ministries. For example, Minvodkhoz served as both a monitor and an enforcer for environmental protection while also acting as the main appropriator (Zile 1971, p. 86). Likewise, the Ministry of Agriculture's main objective was to increase agricultural production and crop yields while also sustaining the fertility of the soils. This created a real perverse form of clientele capture, making environmental protection subordinate to the overall goal of economic development. Neither of these ministries possessed any compelling reason to monitor and enforce violations, since they were largely policing themselves.

It was only in 1987 that the Soviet authorities created the State Committee for Nature Protection (Goskompriroda) to carry out environmental protection in lieu of just environmental monitoring, which meant data collection (Ziegler 1987). As much as Goskompriroda was supposed to promote the rational use of resources and enforce environmental regulations, from its inception it encountered direct opposition from the ministries.[35] Even though the Soviet Union had some of the most stringent regulations on the books, the Soviet authorities never enforced them since it was cheaper for the ministries and farms to pay the fines that were minimal.

The origins of the Aral Sea crisis derive from these agricultural and industrial policies that disregarded social and environmental costs. Soviet planners sacrificed the environment to build grandiose projects symbolizing the triumph of socialism over capitalism. In Central Asia, Soviet planners preferred to build large dams, reservoirs, and canals to harness the water for irrigation. Owing to the vastness of the Soviet territory, they viewed natural resources as inexhaustible. Accordingly, Soviet planners considered water to be an unlimited resource. Along with the perception of plenty, they believed that technological advances would overcome any limitation (Gustafson 1989). Even when faced with increasing water shortages in Central Asia in the mid 1980s, the solution most favored by local scientists and party apparatchiki was to divert Siberia's Ob and the Irtysh Rivers to Central Asia via a huge canal across Kazakhstan.

Cotton Monoculture as a System of Social Control

In Central Asia, to understand water is to understand cotton. As noted above, the Soviet authorities in the aftermath of the Bolshevik Revolution restructured local agrarian and water institutions to forge an economic system predicated on regional specialization. As a result of these policies, the political economy of Soviet Central Asia soon became synonymous with cotton monoculture; yet in order to gain the compliance of the indigenous population, the authorities concurrently needed to invent a mythology of "king cotton" in which cotton was equated with the "blood of life" for the Central Asians (Reznichenko 1992, p. 23). The symbol of cotton was omnipresent. In Toshkent, even the streetlights resembled cotton bulbs, and the main metro station and stadium were named for the cotton worker (Pakhtakhkor). Cotton dominated all discussion of local economic development in the media during the harvesting season, when collective or state farms were either rewarded or censured for how much cotton had been picked. In general, it was impossible to escape the image of cotton and its subsequent effects on shaping state-society relations in Central Asia.

Owing to the all-encompassing role for cotton, many Western specialists framed the Central Asian experience under the Soviet system as one of colonialism in which Moscow's cotton policies directly led to the exploitation of the indigenous population and natural environment in the Aral basin.[36] In fact, cotton monoculture did exact a particularly heavy toll on the populations downstream in the delta—especially the Karakalpaks, who experienced the severest environmental and health consequences from upstream policies that subsidized water use for cotton cultivation. The institutional incentive structure of centralized planning undergirding the Soviet economy generated many of these environmental externalities, since the system did not value any of these natural resources in their own right. Soviet planners priced water artificially low. Indeed, farmers, like Minvodkhoz, received it for free. However, assuming that Central Asia was merely a cotton colony of Russia does not fully capture the intricate web of beneficiaries and losers, and it does not help to explain why so many actors within Central Asia continued to have a vested interest in perpetuating the system of cotton monoculture even in the post-Soviet period.

Although this obviously was a situation in which the hegemon (Moscow) was both coercive and exploitative, it also supplied a public good in the form of stability via an incredibly elaborate patronage system. This is astounding in view of the fact that by the 1980s the economy was not growing; instead aggregate yields were declining as more and more lands were being brought into cultivation in order to meet production targets. Yet, at the same time, the region was characterized by levels of political and social order much higher than found in the Slavic parts of the Soviet Union.[37] The rents extracted from the sale of cotton supported a system of patronage in Central Asia that, in turn, served as a basis for a system of social control.[38] The struggle between the center and the republics was not only about water allocations but also about who was going to receive the bulk of the cotton rents.[39] Thus, in order to assess why rapid cooperation took place immediately after the breakup of the Soviet Union, we need to understand the importance of cotton as a mechanism for social and political control in Central Asia and subsequently as a major constraint on the bargaining positions for water sharing in the post-Soviet period.

The system of social control manifested itself through a layered patron-client network composed of three main interest groups linked together in various reciprocal relationships through the production and delivery of cotton. The first relationship binds Moscow to the republican leadership, including the regional elites. The second relationship integrally connects the republican and regional elites with the general population. The third relationship ties Moscow directly to the population. The extraction of cotton rents provides the basis for these reciprocal relationships. Concerning the first level, Moscow needed cheap cotton extraction. The leaders of the various republics supplied this through the regional leaders in exchange for a certain degree of freedom over internal affairs. Here the regional leaders in particular benefited. As long as they "brought in" good cotton harvests, they could control strong patronage networks through the resources they managed. This leads into the second reciprocal relationship between the republican leaders and regional leaders with the population, which organizationally took place through the collective and state farms that grew cotton. Through the patronage networks they commanded, they could "pay off" the population and gain compliance. Finally, Moscow and the general population also maintained a reciprocal relationship in which Moscow provided social protection and, in exchange, the population did not engage in overt protest. Figure 4.1 illustrates these relationships.

Even Minvodkhoz's existence was entirely dependent on the growth of the cotton sector and bound up in these reciprocal relationships. As annual cotton quotas increased, more land continuously needed to be brought into cultivation, requiring huge investments in new irrigation schemes. On one level, then, Minvodkhoz required information in order to undertake these new irrigation schemes, which meant it needed

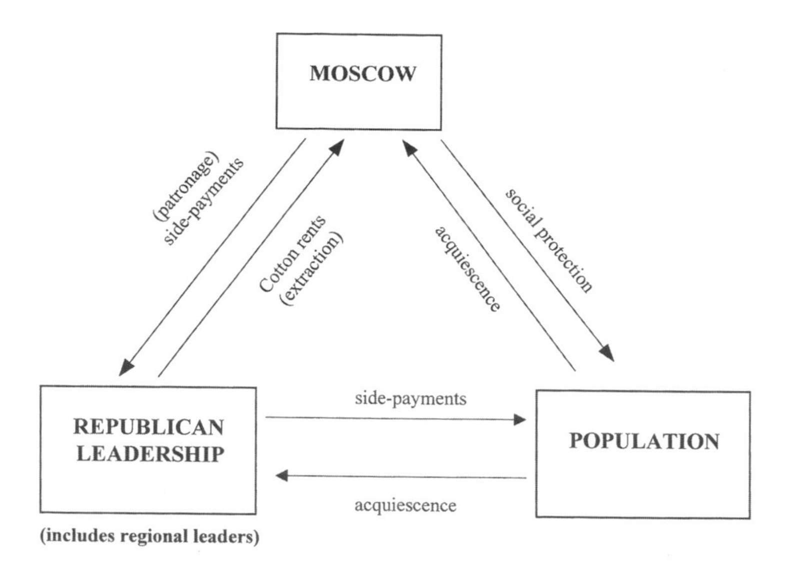

Figure 4.1
The system of social control.

to depend on information collected by the local and regional irrigation administrations. At times, this created asymmetries of information between the center in Moscow and the regional and local levels. Yet, on another level, the republics and the local administrations needed the transfers and subsidies from Minvodkhoz. Here, the All-Union Minvodkhoz could mitigate potential conflicts among the republics over water sharing and internally between water users through the provision of subsidies and transfers in the form of huge infrastructure projects. For example, Uzbek Minvodkhoz used these subsidies to not only expand the irrigation networks but also to build factories, parks, and entire towns. (Several factories on the Chirchik River belonged to the ministry.[40]) Indeed, Minvodkhoz was an immense bureaucracy that in addition to construction projects was responsible for the social services of its employees living in these towns; it operated the hospitals, the schools, and the main recreational facilities. In general, the ministries were the means through which social services were dispensed, and the more large-scale projects that the ministries obtained, the more resources they themselves could distribute to their employees.

In regard to the first reciprocal relationship between Moscow and the republican and regional elites, cotton monoculture, in practice, created a system of personal rule by which a patron (the Communist Party) dispensed resources through its clients (the regional political elite). Owing to the system of economic regional specialization, the system of patronage required a strong state role in agriculture. Moscow acted as a monopsony in which all cotton procurements went directly to the center; Moscow extracted most of the revenue for itself by paying a local price for cotton and then selling it on world markets. In order to ensure that production goals were met, Moscow, however, needed to rely on the republican leaders, who in turn needed to depend on the oblast (regional) leaders, who then had to entrust the rayon (local) leaders to procure the cotton from the heads of farms. One of the ways in which patronage played out in Central Asia was that the predominance of cotton monoculture or regional economic specialization allowed the republican leaders to control many of the main posts and/or political and economic appointments within the republics. Political elites and even farm heads were awarded with promotions and bonuses for "bringing in" the cotton harvest. The implications of this were that the republican leaders could penetrate and affect societal relations and establish firm social control over a large segment of society in ways that the center in Moscow could not by giving preferential access to resources to many local indigenous clients.[41] Simply put, the Central Asian nomenklatura had a stake in perpetuating the system of cotton monoculture as they gained access to special privileges that evaded the majority of the population, primarily better opportunities for social and political mobility.

Indeed, the indigenous population dominated the agricultural sector not only because of cultural factors embedded in a long history of settled agriculture but also for reasons tied to the flow of patronage. Overall, agriculture employed the largest segment of the population in Central Asia. In 1987, the percentage of the population employed in agriculture were the following: Kazakhstan, 23 percent; Kyrgyzstan, 34 percent; Tajikistan, 42 percent; Turkmenistan, 41 percent; Uzbekistan, 38 percent.[42] One of the main reasons underlying the expansion of irrigated cropland in Central Asia was to guarantee employment for the indigenous population, especially after it was shown that they had a low propensity for outmigration. The birth rate in Central Asia was one of the highest in the Soviet Union, and any attempts at cutting back on cotton production could have dramatically exacerbated already high unemployment. The growing of cotton is extremely labor intensive, and in Central Asia most of the harvesting was done by hand rather than with machinery. Gregory Gleason (1990a, p. 85) suggests that the mechanization of agriculture was delayed in Central Asia precisely to avoid increasing labor unemployment and to ensure that people would stay on the farms rather than migrate to the cities. This echoes earlier work (Lubin 1984) that showed that the Central Asians preferred to work in the agricultural sector rather than the industrial sector because it allowed them to accrue benefits from the informal economy even when it appeared that they were unemployed.

Consider here the reciprocal relationship between the republican and regional leaders with the general population. At the real micro level, on the farms, the transfer of patronage and resources manifested itself in yet another form. In particular, the republican leaders were able to ensure political and social order and to garner laborers for the cotton harvest by allowing the farmers to concurrently work in the informal economy. Even when most agriculture was geared toward cotton, families still had private plots where they could produce their own food along with additional crops for sale in the markets. Similar to the ministries, the farmers received subsidized farm inputs such as free water and cheap fertilizers and fuel, which also created disincentives to conserve water. These types of subsidized farm inputs from the center compensated the farmers for the low returns on their output from picking cotton. Instead, they used these practically free inputs for personal purposes in the informal economy, which had spawned various "cottage industries." For example, several households on the Kelesk state farm in Kazakhstan had converted rooms in their houses to raise silkworms.[43] The republican and regional elites essentially turned their backs on "violations" such as these cottage industries in exchange for bringing in the cotton harvest and, moreover, for not protesting about their women and children having pesticides sprayed all over them when they were sent out to the fields to glean the cotton.

Moscow and the general population also had a reciprocal relationship by which the provision of social protection ensured compliance and acquiescence among the indigenous population. Most of the social protection or the provision for "basic needs" came from transfers and subsidies from the center in Moscow to the Central Asian republics, often related to the agricultural sector. The World Bank (1993b, p. 106) reported that in Uzbekistan "the costs of social protection increased inexorably through the 1980s." In 1980 the cost of social protection was roughly 7.4 percent of the GDP; by 1990 it had risen to about 12 percent; by 1991 it had reached almost 29 percent. Moreover, the World Bank (ibid.) noted that "in 1985 and 1990, transfers (from Moscow) covered roughly the cost of the social safety net. In this sense, the system of social protection under socialism was free in Uzbekistan." A similar situation existed in Turkmenistan, where transfers from the union budget accounted for up to 20 percent of total revenues, and most of this supported the social sector (World Bank 1994, pp. 10, 12). It should be emphasized that these two countries were the largest cotton producers in Central Asia, and thus, even when the cotton-producing republics protested to Moscow about the perils of cotton monoculture during the late 1980s, the republican leaders were seeking to acquire increased transfers from the center and a greater share of the revenue from the sale of cotton to be used at their individual discretion along with seeking recourse to mitigate the effects of the Aral Sea crisis.

Such a system of patronage usually tends toward corruption. In the mid 1980s it became apparent that most of the political elite in Uzbekistan had been filling their own coffers by charging the center for cotton that was never produced. The 9 million tons of cotton reportedly delivered in the 1980s turned out to be a massive fabrication (Lipovsky 1995, p. 538).[44] This can partially be attributed to the Soviet incentive structure, which valued quantity over quality. The republican leaders were judged by whether or not they could fulfill the plan and meet the set quota for that year. In Uzbekistan, First Secretary Sharaf Rashidov[45] was held personally responsible for delivering a certain amount of cotton per year, and in the early 1980s he agreed to produce an unrealistic quota of 6 million tons of cotton in one year (Carlisle 1991, p. 33). Even when cotton yields actually declined, the Uzbekistan clients falsified production data to keep up with the center's demands for more cotton. In order to meet these targets, the heads of the collective and state farms simply bribed the inspectors at the procurement centers to inflate the amount delivered (Rumer 1989, pp. 69–70). This was one way for everyone to get a higher price for cotton than what Moscow was officially paying them (which was clearly below market prices).

This padding of data came to light in the mid 1980s when the cotton scandal in Central Asia broke out, involving the collusion of thousands of people among the power elites of Uzbekistan and some close relatives of Leonid Brezhnev in the center.[46] Uzbekistan, in particular, was swept by a massive wave of arrests in which a large percentage of the party elite were accused of corruption in the guise of nationalism (Gleason 1990b). Some argue that Moscow's attack on the party elites was really an attempt by the center to regain control over local party recruitment and to break down the patronage networks that sustained Central Asian state-society relations (Critchlow 1991b).

What Moscow did not realize when it sought to supersede the local authority of the mirab and the aqsaqal through the creation of the state and collective farms and the centralization of water management was that the local regional leaders would use the system of cotton monoculture and regional specialization to consolidate their own local power bases. The system created by Soviet planners was not supposed to reinforce patron-client relations. In spite of this, Soviet institutions usurped local patron-client relations through a process of institutional overlay. For example, the role of the hydrotechnician (vodnik) on the kolkhoz was really just grafted onto the previous traditional role of the mirab. Furthermore, the republican leaders used the patronage networks and resource transfers to counterbalance the various regions within the republics in order to maintain social and political control, even in the face of a mounting economic and environmental crisis.

5

The Need for Aid: Failed Reform, Potential Conflict, and the Legacy of Cotton Monoculture

In the early to mid 1980s, once Moscow realized that the Central Asian leaders had used their control over cotton production to dominate local party recruitment, it initiated a sweeping anti-corruption campaign to break down the system of political control in Central Asia. This anti-corruption campaign was a part of General Secretary Mikhail Gorbachev's larger reform process (1985–1991), which aimed to revitalize the Soviet economy through a program of domestic economic restructuring (perestroika) and political liberalization (glasnost). The coincidence of glasnost's effects on Central Asian domestic politics and the growing international awareness of the Central Asian environmental crisis laid the groundwork for international intervention in the post-Soviet period. In particular, the unintended consequences of Gorbachev's political and economic reforms led to a sharp rise in eco-nationalist movements and in small-scale ethnic conflicts tied to scarce resources. Under growing pressure from domestic interest groups and international organizations, the Soviet leaders sought an all-Union solution to rectify the mismanagement of water policies in Central Asia.

The changing domestic context during the Gorbachev period and the breakup of the Union created a demand for outside assistance right after independence. Specifically, aid was needed to compensate many of the groups directly affected by the Aral Sea crisis and to offset the new asymmetries of interests and of capabilities that emerged with the unsettling of physical and political boundaries. Whereas glasnost engendered internal domestic discontent and incited small-scale ethnic conflicts within the republics, independence created a real potential for conflict among the newly independent states as the water sector began to compete with other sectors, such as energy.

Yet, during the first year after independence, the Central Asian successor states retained the Soviet arrangements for water management in the Aral basin. These states preferred a policy of inertia to institutional change. However, the Central Asian predisposition toward inertia could not explain adequately why the Central Asian states agreed to the subsequent creation of new institutions. Rather, the 1992 agreement to maintain past practices was precarious at best because of the need for assistance to pay the costs of the transition and to address the environmental and economic consequences of the Aral Sea crisis. In short, the need to fill the void left by the demise of the Soviet Union forced the Central Asian states to seek international intervention and to constitute new interstate water institutions.

Confronting the Aral Sea Crisis: How Glasnost Affected Domestic Politics

The acute awareness inside and outside the former Soviet Union of the desiccation of the Aral Sea in the early 1980s coincided with indications that the entire Soviet regime was facing a severe economic and political crisis. Economic growth had fallen sharply, and these dwindling growth rates were associated with the blatant misuse of natural resources along with the technological backwardness of the Soviet economy. By the mid 1980s, the Soviet authorities could no longer disregard earlier warnings from the scientific community about the environmental repercussions of the indiscriminate use of water for irrigation compounded by a lack of lined canals and adequate drainage in Central Asia.[1] The desiccation of the Aral Sea was visible to the naked eye: fishing boats were stranded in the sand of the exposed seabed.

The Soviet and republican leaders had willingly sacrificed the Aral, believing the economic benefits of increased agricultural production to be worth the environmental tradeoff. Typifying this view, the First Deputy Prime Minister of Minvodkhoz, Polad Polad-Zade, brazenly remarked that "the Aral should die beautifully."[2] Yet, in response to the impending economic crisis in the Soviet Union, members of the Soviet leadership began to consider various ways to end more than 10 years of economic stagnation. Gorbachev had formulated the beginnings of a program to revitalize the Soviet economy by December 1984. Soon after he became General Secretary of the Communist Party, in March 1985, he initiated a series of policies to restructure the ailing economy and better the living standards of the population.[3]

As a central component of perestroika, the newly appointed Soviet reformers sought to curtail past practices that prioritized economic growth over quality of life issues.[4] Previously, Soviet planners flagrantly overlooked human health and environmental issues to ensure that production targets were met, ignoring the environmental and social costs of cotton monoculture in Central Asia. But as cotton production declined, Soviet planners began to encourage environmental protection as a means to improve economic efficiency.[5]

In response to the worsening environmental and economic crises, in January 1988 the Soviet leaders created the State Committee for Nature Protection (Goskompriroda) to oversee environmental protection, replacing the State Committee for Hydrometeorology and the Environment (Gidromet), which had previously served as the main environmental agency.[6] Contrary to the Soviet Union's emphasis on centralized decision making, Goskompriroda's responsibilities were decentralized in order to cope with the widespread and pervasive environmental degradation throughout the Soviet Union. Though it retained a small staff in Moscow, most of its activities were channeled through parallel nature protection committees in each of the republics, oblasts, and rayons. Many of Goskompriroda's attempts at mitigating environmental problems were not just for the sake of the environment alone but ostensibly were intended for overhauling the centrally planned economy.[7]

The Soviet authorities sought to counter the intransigence and wastefulness that decades of centralized planning had generated by enabling the local offices to fine violators and then to utilize this revenue to support their operations. This was part of the center's new emphasis on republican self-reliance. From the onset, the republican and local offices of Goskompriroda encountered direct opposition from the very ministries they were supposed to be regulating. Dr. Vladimir Gregorovich Konyukhov, Vice-Chairman of Goskompriroda in Uzbekistan, pointed out that, although Goskompriroda's authority and capacity were at their strongest under glasnost, his republic's office was only able to shut down one chemical plant in the city of Kokand for groundwater contamination in 1989.[8]

The Rise of Environmental Activism and "Eco-Nationalism"

To expedite economic reform, Gorbachev began to promote glasnost (the opening of the political sphere to dissent). In the environmental sphere, glasnost was intended to help Goskompriroda and the regional and local elites disseminate information about the relationship between environmental and economic problems.[9] Glasnost, however, enabled individuals to voice their demands and grievances without fear of reprisal from central and regional government officials. Yet glasnost was not intended to internally divide the country or to fuel separatist movements. According to Goldman (1992, p. 4), it was only supposed to "make the Soviet Union a more effective political and economic force" through wiping out political corruption and abuse and combating industrial inefficiency. Nevertheless, the integral relationship between economic and ecological problems spawned nationalist movements in many regions of the Soviet Union. Glasnost enabled grassroots movements to develop environmental platforms with nationalist overtones (Dawson 1996). In fact, environmental issues served as a focal point for political mobilization throughout the Soviet Union, especially as the population became more cognizant of the dramatic rise in health and environmental problems that was attributable to Soviet economic policies (Peterson 1993, pp. 193–234).

Glasnost forced the Soviet leaders to declassify many environmental disasters, and the nuclear accident at Chernobyl in 1986 provided the first real test of this policy of openness.[10] Delayed though it was, the disclosure to the international community that an accident had occurred signaled to the domestic community that secrecy would no longer be tolerated. Chernobyl showed Soviet citizens not only that they were at risk from nuclear contamination but also that Moscow had often overlooked their social and health interests in the pursuit of industrialization and modernization. Glasnost provided a fertile ground for many Central Asians to conjecture that their poor environmental and health situation was related to the policy of regional economic specialization of cotton monoculture.

As a result of the new openness, many Central Asian writers and intellectuals began to express their resentment toward the omnipresence of cotton monoculture in Central Asian society. They published numerous articles in the national and republican newspapers on the sharp rise in infant and maternal mortality, and they documented the poor working conditions women and children experienced during the cotton-picking season.[11] Several women on the Sovkhoz Savai in the Kurgan Tipenski region in the Fergana Valley intimated that for decades they had been silently wondering about the health consequences of having defoliants sprayed on them as they worked in the cotton fields.[12] These women had noticed a rise in illness, but were unsure about the causal links, and only after the articles began to appear did they fully start to grasp and identify the relationship between human health risks and Soviet policies for growing cotton monoculture. Glasnost, likewise, enabled many Russian writers and intellectuals to make the plight of the Aral Sea known to a broader audience. For instance, Grigori Reznichenko of the literary journal *Novy mir* led an expedition in Central Asia, "Aral-88," to inform its readership of the demise of the Aral Sea.[13]

Alongside the explosion of newspaper articles, a loosely organized opposition was emerging. Following the example of environmental movements in which Russian national writers spearheaded the fight to save Lake Baikal and prevent the Siberian water diversions, one of the first grassroots movements to form in Central Asia was the Committee for Saving the Aral Sea, led by the writer Pirmat Shermukhamedov.[14] The topic of the environment enabled many Central Asian writers and intellectuals to promote national issues that had been festering for sometime. Shermukhamedov and other Central Asian intellectuals rallied around the Aral Sea as a proxy for long-suppressed questions of cultural survival and regional self-determination. The writer and head of the Uzbek Writers' Union, Adyl Yakubov, in a speech delivered at a plenum of the USSR Writers' Union in 1988, argued that the desiccation of the Aral Sea and the surrounding health crisis in the Near-Aral Region were indubitably tied to the economic and social system of cotton monoculture. In his speech, he pleaded for Moscow to reduce Uzbekistan's cotton quota in order to improve not only the health and ecological situation in Central Asia but also the economic conditions of the Central Asian peoples.[15]

Likewise, Tulepbergen Kaipbergenov equated the desiccation of the Aral Sea with the Chernobyl accident.[16]

Nascent nationalist movements, such as Birlik (Unity) and Erk (Freedom), also used the question of the desiccation of the Aral Sea in their struggle against Russian dominance to press for cultural autonomy in issues such as language concerns. These nationalist movements considered cotton monoculture to be the manifestation of Soviet exploitation and the lack of control the Central Asians had over their own destiny. The Central Asians saw themselves as living in a "cotton colony" in which they were slaves to the directives of Moscow (Critchlow 1991a, p. xii). Moreover, the "cotton affair" and the subsequent purge of the indigenous republican elites only heightened the tension between the republics and the center. Yakubov, in the above-mentioned speech, captured the sentiment of many Central Asians who believed they were twice victimized by a policy generated in Moscow that forced them to grow only cotton and then have to falsify the data to keep up with the increasing demands from Moscow.

Overall, those with nationalist aspirations or other grievances against Moscow mobilized around the Aral Sea crisis. Like movements in other republics and regions that were convinced that environmental ills afflicting their territories were due to Moscow-driven policies, the Central Asian movements couched their environmental arguments in terms of colonialism and exploitation. As these groups in Central Asia developed their platforms, they began to call for the devolution of the power to make decisions to the republican and regional levels as a means to strengthen local and cultural autonomy. This created demands for regional control and for the ability to make decisions about their natural resources.

In short, the Aral Sea crisis accentuated the growing tension between Moscow and the Central Asian republics. In the late 1980s, leaders of Central Asian republics began to champion the claims levied by the opposition movements against Moscow and to adopt the language of the nationalist movements. For example, President Islam Karimov attributed Uzbekistan's dire economic situation to the many years of centralized economic policies that had turned the republic into a "raw-materials appendage" that received unfairly low prices for its cotton.[17] As of December 1990, all the Central Asian republics had declared sovereignty, asserting their rights to control land, water, and other natural resources within their respective territories. Yet the Central Asian leaders endorsed these steps not to gain full independence from the Soviet Union, but rather to press the center in Moscow for more regional autonomy as a means to strengthen their own authority within their respective republics. Simply put, the Central Asian leaders hoped to gain greater control over the cotton harvest in order to recover more of the hard currency from sales abroad.

The Upsurge of Local Conflicts

The growing tension between the center and the periphery over the water crisis was not the only issue over which disputes were manifesting themselves. Both the titular leaders of the republics and the authorities in Moscow faced an unprecedented rise in intra-republic conflicts in which ethnic groups were competing for control and access to scarce resources. Parallel to the growth in eco-nationalist movements, there was an upsurge in small-scale ethnic conflicts in Central Asia during the Gorbachev period. Many of these incidents involved water and land issues (Klötzli 1994). For example, in 1989, Tajiks and Uzbeks quarreled over land and water rights in the Vakhsh Valley. According to local press accounts, increased water logging of the soil as a result of the poor irrigation networks and drainage system forced the resettlement of a group of Tajiks to another sector of the same kolkhoz, which required the redistribution of land. The Uzbek inhabitants, whose personal plots would be lost, vehemently opposed the transfer of agricultural land.[18] For more than 2 weeks in early June 1989 there was ethnic strife between ethnic Uzbeks and Meskhetian Turks in the Fergana Valley. A large number of Meskhetian Turks were evacuated from Uzbekistan to prevent further bloodshed.[19] Estimates suggest that at least several hundred Meskhetian Turks were killed in these riots.

In June 1989, a long-standing dispute between Tajiks and Kyrgyz came to a head over land and water rights in the Isfara-Batken district along the border between the two republics (Atkin 1993, p. 372). Plans to build an interrepublic canal sparked the dispute when the Kyrgyz perceived that they would not receive sufficient water from the Tajiks to irrigate all their fields.[20] In June 1990, violent conflict broke out between Kyrgyz and Uzbeks in Osh, Kyrgyzstan, resulting in several hundred deaths. The immediate cause for the violence was the official permission to reassign to local Kyrgyz the land of a collective farm that Uzbeks had been farming in order to build residential housing. The riots were halted only with the imposition of a state of emergency and the intervention of Soviet troops.[21]

Overall, both environmental and ethnic issues had become increasingly salient during the late 1980s, and they often were so closely intertwined that it was difficult to distinguish which one actually triggered the conflict. Nunn, Lubin, and Rubin (1999) suggest that "the ethnic part of the conflicts generally reflected the fault lines along which the crises erupted, but not the underlying causes." As both the economic and environmental situations continued to deteriorate, the Central Asian leaders, moreover, found themselves ill-prepared to deal with this rise in small-scale ethnic conflicts. These outbursts of small-scale violent ethnic conflict, coupled with the rise in eco-nationalist movements, only fortified the Central Asian republican leaders' resolve to seek solutions from Moscow to mitigate the water crisis. They concurrently sought to reduce the burden of cotton production on the Central Asian republics as a means of diffusing the growing discontent within the republics.

All-Union Solutions to the Aral Sea Crisis

Although the Central Asians demanded more local control over internal decision making, this did not preclude them from seeking an all-Union solution to the water crisis. Rather than pursue a strategy to gain complete independence, they were only asking for greater autonomy and a quick response to the Aral Sea crisis within the confines of the Soviet system. Indeed, the Central Asian republics were depending on Moscow to compensate them for the environmental crisis created by Soviet economic policies. The plan that initially received the most attention was the Sibaral project, which called for diverting the Siberian rivers to replenish the Aral Sea (Micklin and Bond 1988).[22] Soviet planners first proposed turning the Northern rivers southward as a part of Stalin's great plan to transform nature in the 1950s, but by 1986 growing opposition in Russia and the exorbitant financial costs of the project led Gorbachev to officially abrogate this option.[23]

Cancellation of the river-diversion plan not only increased the rift that was developing between the Central Asian republics and Russia but also pitted region against region. In addition to viewing themselves as cotton slaves to Moscow, the Central Asians had reason to believe that Moscow was discriminating against them in favor of the Russian nationalists in Siberia. Moscow erred in its estimation of how the cancellation of the river diversion project would be perceived within Central Asia. From the Central Asians' viewpoint, their only chance at economic revitalization depended on the procurement of additional water from outside the region, whereas Moscow saw improvements in water efficiency as a more appropriate solution. Nevertheless, in 1988, in response to these increasing attacks on Moscow, especially those demanding the diversification of agriculture in the region, the Politburo called for a modest reduction in the cotton quota and for additional water-conservation measures (Critchlow 1991a, p. 73).

Near the end of the glasnost period, the Central Asian leaders became even more assertive in their demands from Moscow. For example, on June 23, 1990, while signing a joint declaration on economic, scientific-technical, and cultural cooperation, they pronounced that the Aral Sea crisis could not be solved by regional efforts alone, and instead appealed to Moscow to declare the Aral region a national disaster zone.[24] In this petition for immediate assistance, they raised the possibility of reinstating the river-diversion project; moreover, they suggested that international organizations be included in the resolution process.[25] Reinstating the river-diversion project or devising an all-Union option would shift the burden of responsibility away from the Central Asian leaders and back to Moscow.

In the fall of 1988, in response to the mounting public pressure, the Central Committee of the Communist Party of the Soviet Union and the all-Soviet Council of Ministers issued a decree on "measures for the radical improvement in the ecological situation and sanitary situation in the Aral region and for raising the effectiveness of use and strengthening the protection of water and land resources in the basin."[26] In order to prevent further deterioration to the Aral Sea ecosystem, the decree guaranteed a minimum inflow into the sea that would incrementally increase from 8.7 km^3 in 1990 to 21 km^3 by 2005.[27] The flow would be increased by improving the efficiency of irrigation, reconstructing old canals, and routing drainage water to the sea.[28] In November 1989 the Supreme Soviet issued a resolution establishing a research and planning framework for rehabilitation efforts in the Aral Sea region and constituting a government commission to carry out this task (Micklin 1992a, p. 86). In 1990 the Government Commission on the Development of Measures for Restoring Ecological Equilibrium in the Aral Region announced a contest for the best ideas for improving the overall situation in the Aral basin and formed an ad hoc group of prominent Soviet scientists to summarize proposals and formulate a final concept that would become official government policy. The State Commission on the Aral issued its final report in 1991, outlining the various concepts developed in the contest. The final report, titled Concept of Conservation and Restoration of the Aral Sea and Normalization of Ecological, Sanitary, Medical, Biological and Socio-Economic Solutions in the Aral Region, included a program to improve land and water use, diversify the economy in the region with a reduction in cotton, improve local health and living standards, and guarantee more inflow to the sea (Levintanus 1992). Implementation of the program was estimated to cost 60 billion (1990) rubles by year 2010, but the Soviet Union broke up and these efforts were never realized or implemented.[29]

How Glasnost Affected the International Community

At the same time glasnost created the conditions for grassroots and opposition movements to form in the Soviet Union, it raised awareness in the West regarding the amount of environmental destruction within the Soviet Union. Disasters such as Chernobyl revealed the precarious state of the environment behind the Iron Curtain. With easier access to the Soviet Union, IOs and Western NGOs forged contacts with Soviet environmental groups and scientists to discuss various environmental issues within the Soviet Union. A growing interest on the part of IOs and Western NGOs in the Aral Sea crisis was emerging, and the inroads made by the international community under glasnost served as the precursor to later international involvement in the post-Soviet period.

Even government officials began to recognize the need to solicit international assistance, both scientific and technical. In a roundtable on ecology, Kakimbek Salykov, Chairman of the USSR Supreme Soviet

Committee on Ecology and Rational Utilization of Natural Resources, told Polad Polad-Zade, First Deputy Minister of Minvodkhoz, that the time had come to turn to the international community for solutions to help extricate the Central Asians from the water crisis, especially since past Soviet solutions were responsible for the current crisis.[30] Then, in January 1990, the United Nations Environmental Program (UNEP) signed an agreement with the Soviet Union for a 2-year program for developing a rehabilitation plan for the Aral Sea region and the near-Aral region. Drawing on the concept developed by the Government Commission, UNEP prepared a diagnostic study based on several fact-finding missions to Central Asia that included foreign and Soviet working groups. This study was originally intended to serve as a basis for an action plan, but the breakup of the Soviet Union abruptly halted the program. Nevertheless, we can consider this UNEP initiative to be the first attempt at multilateral involvement in the Aral basin. At that time, UNEP only had to negotiate with the Soviet government and not five independent governments, and the UNEP program was concerned more with technical, economic, and scientific solutions rather than with political solutions to the Aral Sea crisis.

While glasnost opened up channels to IOs, it also provided new opportunities for Western NGOs to interact with nascent Soviet NGOs. By the end of the Soviet period, the Socio-Ecological Union in Moscow, an umbrella organization for many environmental NGOs in the Soviet Union, began to coordinate activities and exchanges with ISAR (then the Institute for Soviet-American Relations).[31] In addition to spurring domestic environmental activism in the late 1980s, glasnost inspired international campaigns to save the Aral Sea and to preserve the cultures of the peoples living near the sea.

The Aral Sea International Committee (based in Sausalito, California) was one of the first groups involved with the Aral Sea issue. Beginning in March 1991, that committee directed its activities toward garnering attention and assistance to the needs of the local populations whose livelihood was directly threatened by deterioration in water quality and the collapse of the fishing industry in the Aral Sea.[32] Its main counterpart was the Union for the Defense of the Aral Sea and Amu Darya, based in Nukus, Karakalpakstan.

Glasnost enabled local NGOs and eco-nationalist movements in Central Asia to establish ties to the West and to raise the Aral Sea crisis in international forums. Local activists used the internationalization of the Aral Sea crisis to combat the legacy of Soviet secrecy and to rouse both Moscow and the Central Asian leaders into taking tangible steps to address the Aral Sea crisis. Many of these newly formed local NGOs and environmental and nationalist movements continued to operate in the post-Soviet context. They continued to remind the Central Asian leaders that they could not overlook the Soviet legacy of environmental mismanagement. In short, Gorbachev's policies of glasnost and perestroika changed the domestic context and increased international awareness of the Aral Sea crisis.

The Collapse of the Soviet Union: Changing Domestic Conditions and Potential Interstate Conflicts

The disintegration of the Soviet Union into its fifteen constituent parts shelved the all-Soviet concept for improving water management, health, and ecological conditions in the Aral Sea region. The constitution of new juridical borders furthermore disrupted previous multilateral efforts to design an action plan for the Soviet Union to deal with the Aral Sea crisis. UNEP curtailed its program when the influence of Moscow-based working groups waned. Instead, the Central Asian states inherited the previous system for water management without either an authority to guide it or the financial resources to maintain it. The water basin agencies (the BVOs) were left with ambiguous responsibility for interstate water distribution. Independence shifted the distribution of capabilities and interests among the water users in the basin, creating new and competing interests in the water sector.

Moscow no longer needed to respond to the demands of the Central Asian leaders; however, it faced new and daunting obligations concerning how to restructure the Russian economic and political system. Disputes of the sort that had been resolved outside the region by the authorities in Moscow were left for the Central Asian republics as challenges for which they needed to devise their own regional solutions. The Central Asian leaders, who before could use the growing discontent within

Central Asia as a bargaining chip with Moscow, found themselves accountable to increasingly restless and boisterous populations. In the spring of 1992, rather than appeal to Moscow for a solution to the Aral Sea crisis, representatives from Karakalpakstan published an open letter, addressed to the five Central Asian presidents, that pleaded for an all-Central Asian solution.[33] They asked for at least 30 km^3 of water that year in order to begin to restore the Aral Sea.[34] This statement was indicative of the shift in perception of the local population as to who now held responsibility for mitigating the Aral Sea crisis.

In place of Moscow, the Central Asian leaders were suddenly bound to provide for their respective populations' immediate basic needs. If they were to do so, it would, however, be with limited resources, since independence had cut off most of their subsidies and resources transfers from Moscow. At the time of independence, all the Central Asian republics were consuming more than they produced and incurring both a domestic and a foreign trade deficit. In large part because of this dependence on Moscow, the Central Asian republics were the last ones to declare their independence for fear of losing these resource transfers, which had already exacerbated the deteriorating economic and living conditions in Central Asia. Independence created a need for outside resources to compensate for the loss of financial transfers from Moscow, which in turn opened the door for international activity and funds to substitute for some of their missing resources. (See table 5.1.)

In addition to disrupting the domestic situation, independence created a new form of potential conflicts at the interstate level. The transformation of the Aral basin into an international river system generated a large

Table 5.1
National income and trade balances for the Central Asian states in 1988 (millions of rubles). Source: Kaiser 1994, pp. 336–338.

	National income balance	Total trade balance
Uzbekistan	−3100	−1841
Kazakhstan	−5597	−7255
Kyrgyzstan	−998	−1149
Tajikistan	−680	−1133
Turkmenistan	−146	−284

number of potential upstream-downstream disputes. Moreover, independence enabled each new successor state to define its own independent strategy for the use of this shared resource system. As with most other international river systems, upstream interests collided with those of the downstream users. A real likelihood for conflict over water existed on two levels at independence (Klötzli 1994). First, there was a potential for conflict among the republics at the interstate level between upstream and downstream riparians related to both water quantity and quality. Second, the tenuous state boundaries had created a potential for internal conflicts among domestic water and land users. These potential conflicts in the Aral basin were a function of the disjuncture between political and physical borders.

A reexamination of the topographical and hydrological factors discussed in chapter 2 reveals numerous potential conflicts of interests and capabilities among the riparians in the Aral basin after independence. The new upstream-downstream dichotomy resulted in a situation typical of most international river systems in which the benefits of cooperation are highly asymmetrical and are distributed unevenly. In post-Soviet Central Asia, the republics that were really water poor, Uzbekistan and Turkmenistan, were also those most dependent on the water resources from outside their states for irrigated crop production. Uzbekistan did not control the sources of the three main rivers, the Syr Darya, the Amu Darya, and the Zarafshon, but used three-fifths of the regional water supplies for irrigated agriculture (Smith 1995, pp. 356–357). Uzbekistan also lacked authority to operate the hydrotechnical installations on the Syr Darya, known as the Naryn Cascade, at will. At the same time, three-fourths of Central Asia's population resided in the midstream and downstream territory of the basin of which Uzbekistan made up over half of the population.

Upstream-Downstream Disputes in the Syr Darya Basin

In the Syr Darya basin there were at least two related areas wherein conflicts could have erupted at the interstate level. The more salient one revolved around the different scenarios for the management of water released from the Toktogul Reservoir. With independence, Kyrgyzstan controlled most of the Naryn River, a tributary of the Syr Darya on which some of the main hydroelectric stations, dams, and reservoirs were located. The Toktogul Reservoir was the largest and the only reservoir with substantial storage capacity. It essentially determined how much water was released to the lower reservoirs along the Naryn cascade. It had a generating capacity of 1200 megawatts and a total reservoir storage capacity of 19.5 km^3, with 14 km^3 available. Soviet planners had constructed it to meet irrigation demands downstream rather than to produce energy upstream. Owing to diminishing energy supplies from Russia and the other Central Asian republics, Kyrgyzstan experienced winter energy shortages. To make up for these, Kyrgyzstan chose to intermittently use the Toktogul power plant to generate electricity. Although the upstream states were rich in water resources, they lacked energy resources, whereas the downstream states possessed vast oil and gas reserves. Thus, when Kyrgyzstan operated the Toktogul hydroelectric station in the winter, the water released was diverted to a local depression, the Arnasai lowland, because of the winter freezing of the lower Syr Darya, and as a consequence the water did not reach the sea.[35] In a dry year, Kyrgyzstan could reduce the water flow to Uzbekistan in the spring and summer, when the demand for irrigation was at its peak downstream.

In contrast to the energy-rich Central Asian states, Kyrgyzstan was mired in a severe economic crisis. The situation was exacerbated by Kyrgyzstan's lack of alternative resources such as oil and gas or even substantial amounts of cotton that could be sold abroad to secure foreign revenue. Almost immediately after independence, many Kyrgyzstanis realized that water was their only valuable asset and that selling hydroelectric energy could provide a much needed source of foreign revenue. The head of the division of Energy and Natural Resources at the Institute for Strategic Studies in Kyrgyzstan proclaimed: "Uzbekistan should pay for water if they want to maintain an irrigation regime. Kyrgyzstan should sell water or at least exchange water for gas."[36] Many water specialists and parliamentarians in Kyrgyzstan echoed this sentiment, especially since Kyrgyzstan after independence had to pay Uzbekistan hard currency for gas.[37] Indeed, during the Soviet period, Kyrgyzstan routinely sold excess power from the hydroelectric stations to Uzbekistan and Kazakhstan in the summer in exchange for irrigation water.[38] Independence disrupted such former interdependencies, since the countries began to ponder divergent and often conflicting plans for the development and use of their water and energy resources. For example, the government of Kyrgyzstan actively pursued foreign assistance to complete two other hydroelectric projects further upstream, Kambarata 1 and 2, which were only partially built by Soviets engineers.[39] Kyrgyzstan also sought export markets for the sale of hydro-electricity to either Pakistan or China.[40]

As the conflict over Toktogul illustrates, upstream and downstream states had competing interests and differing capabilities. Yet this was not the only potential upstream-downstream conflict in the Syr Darya basin. Whereas upstream use affected the quantity of water delivered downstream, it also determined the quality of the water. In short, downstream users faced a different set of constraints. This was particularly true for the midstream agricultural users in the Fergana Valley (primarily Uzbekistan) and the Golodnaya Steppe (Uzbekistan) and for the downstream users in Shymkent (Kazakhstan) and Qyzlorda (Kazakhstan). All these users wanted to ensure ample water supplies from Kyrgyzstan. At the same time, the midstream users' interests diverged from those of users further downstream. The midstream users needed water primarily for agriculture while the downstream users in Kazakhstan were concerned about preventing further shrinking of the sea and about procuring clean water for drinking as the water they received was latent with agricultural runoff.

Upstream-Downstream Disputes in the Amu Darya Basin

The Toktogul Reservoir is only one example of how the Soviet Union's collapse politicized the control and the use of hydrotechnical assets. Similar conflicts could have transpired in the Amu Darya basin between the upstream and downstream states, but the civil war in Tajikistan precluded such actions as in the Syr Darya basin.[41] Of much greater concern in the upstream reaches of the Amu Darya basin was the issue of dam safety. Many of the dams were in want of serious repair. Several had already collapsed, which led to flooding and unregulated releases of the water.[42]

In the Amu Darya basin the demarcation of new political borders elevated the status of various domestic canals such as the Kara Kum Canal to the international level. Uzbekistan and Turkmenistan divided the Amu

Darya water at Termez equally between them, but this past allocation could have become an international point of contention since Turkmenistan had expressed its intentions to extend the canal so that it could add an additional million hectares of cultivated land.[43] These two midstream states both had a clear interest in procuring water for the production of cotton, which provided a disproportionate share of their primary revenue. Furthermore, Dashhowuz province (Turkmenistan), Khorazm province (Uzbekistan), and the Autonomous Republic of Karakalpakstan (Uzbekistan) shared the lower Amu Darya, another important agricultural region. Here, problems arose over a canal that Turkmenistan was building away from the Tuyamuyun Reservoir to improve water delivery for the oasis in the lower reaches of the Amu Darya and to increase its control over the water supply within Turkmenistan.[44] Yet the Tuyamuyun Reservoir, built by Soviet planners to increase Uzbekistan's storage capacity, was under Uzbekistan's jurisdiction.

The clash between upstream and midstream interests for agriculture and the downstream interests for clean potable water represented a situation in which the disincentives for cooperation were starkest. The downstream users in Karakalpakstan possessed little if any bargaining leverage over the upstream users, since they lacked coveted economic resources desired by the upstream users. In the Syr Darya basin, by contrast, energy was exchanged for water. Still worse was the situation of downstream users in Karakalpakstan, who were located at the end of the flow of the river and who thus had no choice but to use the contaminated water filled with waste and effluents for drinking purposes. Table 5.2 summarizes some of the major potential conflicts in the Aral basin.

A Related Potential Water Conflict: The Fergana Valley

The situation in the Fergana Valley captures the way in which resource issues are intertwined with the broader issues of ethnicity, economic development, and state formation in the post-Soviet context. With independence, territorial borders assumed particular significance for the Central Asian states, as they were only constructed in the Soviet period. Although the Central Asian governments maintained their inherited borders, these borders were particularly sensitive for the three countries that shared the Fergana Valley—southern Kyrgyzstan, northern Tajikistan and eastern

Table 5.2
Potential water-related conflicts in Aral basin. (The Aral basin comprises several different water basins. Here I have disaggregated several of the basins in order to show the variation in types of conflicts of interests and capabilities.)

Drainage basin	States that share basin	Potential conflicts	Nature of conflict
Syr Darya	Kyrgyzstan (upstream), Uzbekistan (midstream), Tajikistan (midstream), Kazakhstan (downstream)	Among Kyrgyzstan, Uzbekistan, and Kazakhstan over water releases from Toktogul Reservoir	Energy vs. irrigation
		Between upstream agricultural users in Fergana Valley and Golodnaya Steppe and downstream users in Kazakhstan	Quantity and quality of water; agriculture vs. potable water
Amu Darya	Tajikistan (upstream), Uzbekistan (midstream), Turkmenistan (downstream), Karakalpakstan in Uzbekistan (downstream)	Between Turkmenistan and Uzbekistan in regard to water withdrawals from Kara Kum Canal	Quantity
		Other potential conflicts over water sharing in lower Amu Darya between Uzbekistan and Turkmenistan	Quantity
		Between upstream users in Tajikistan and downstream users in Turkmenistan, Uzbekistan, and Karakalpakstan	Quantity and quality; long-term potential conflict between energy vs. irrigation
Zarafshon	Tajikistan (upstream), Uzbekistan (downstream)	Between upstream users in Tajikistan and downstream users in Uzbekistan	Quantity

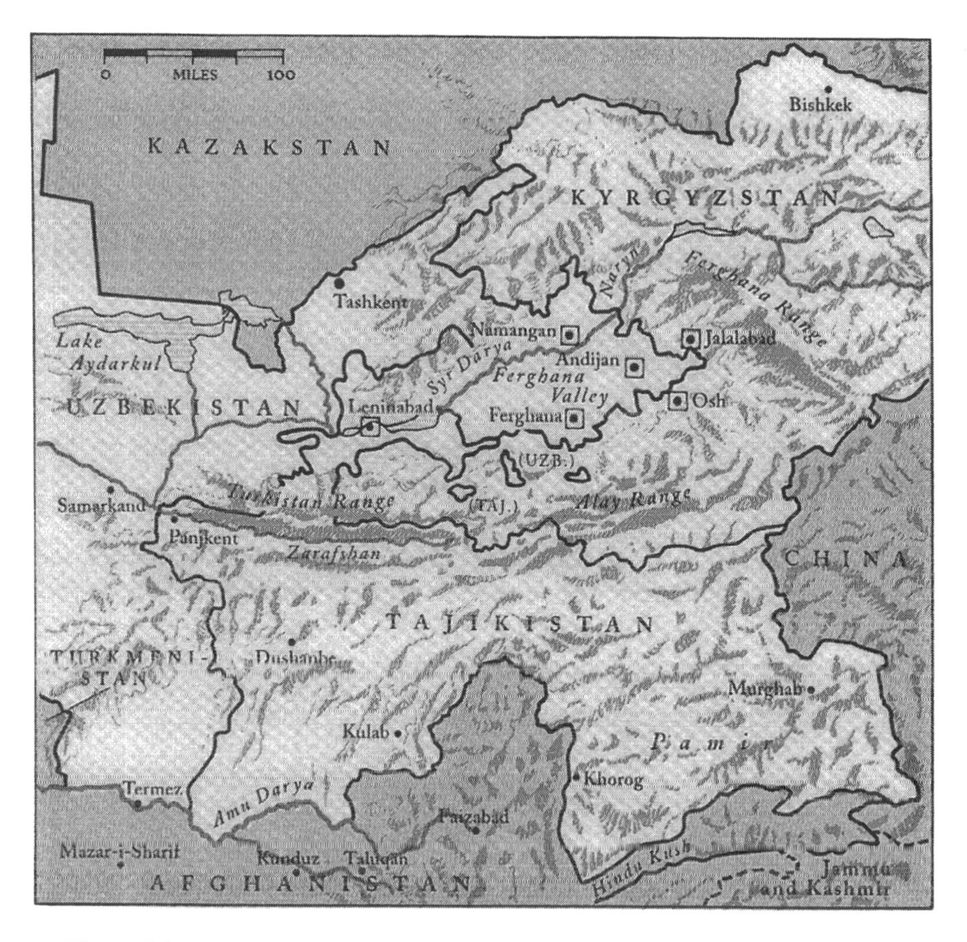

Figure 5.1
The Fergana Valley. Source: Nunn, Lubin, and Rubin 1999.

Uzbekistan—because they cut across ethnic groups and across agricultural regions. (See figure 5.1.)

Although the Fergana Valley covered only 5 percent of post-Soviet Central Asia, it was home to 20 percent of the region's people (more than 10 million people). Forty-five percent of the irrigation area of the Syr Darya basin was located within the Fergana Valley. As a result of independence, canals built to support agriculture throughout the Fergana Valley transcended these new political jurisdictions. This territory contained some of the most vital and productive irrigated areas such as Jalal-Abad and Osh in Kyrgyzstan; Andijon, Namangan, and Fergana in Uzbekistan, and Leninabad (Khujand) in Tajikistan, which all relied on the Syr Darya and its tributaries for irrigation. The competition for scarce resources among highly intermingled ethnic groups in the Fergana Valley affected interstate relations in the post-Soviet context. Just consider the above-mentioned conflict between Tajiks and Kyrgyz in the Isfara-Batken region over irrigation water. During the Soviet period this conflict was internal and localized, but after independence these micro-level conflicts had international ramifications as a result of the importance of new national identities associated with the formation of statehood. Since micro-level conflicts could reinforce both micro and macro identities, subnational level conflicts could turn into interstate conflicts—for example, along the new international border in what previously had been the two neighboring districts of Isfara in Tajikistan and Batken in Kyrgyzstan.[45]

The predominance of ethnic Uzbeks in the Fergana Valley contributed to the tension among ethnic groups. Close to three-fourths of the population in the Fergana Valley was ethnic Uzbek. In Osh, as of 1996, ethnic Uzbeks were the largest group—about 40 percent of the population (Nunn, Lubin, and Rubin 1999). Waters that originated upstream in Kyrgyzstan flowed through the Fergana Valley. Although Kyrgyzstan expressed a firm interest in promoting hydroelectric energy, it needed to be incredibly sensitive to the socio-economic situation in the Fergana Valley. Most of its agricultural base outside of animal husbandry came from its part of the valley and was also tied to cotton monoculture. If Kyrgyzstan reduced the water flow during the summer months, this would exacerbate tensions between ethnic Uzbeks and ethnic Kyrgyz in its two southern oblasts, Osh and Jalal-Abad, as they would have to compete for an even scarcer resource.

Inertia and Institutional Continuity

The above discussion enumerated the ways in which competing interests and capabilities began to form in the late Gorbachev period and intensified with the introduction of new borders after independence. When the Central Asian republics unexpectedly gained their full independence in

December 1991, it remained unclear whether or not they would continue to share the water resources of the Aral basin as newly independent states. The superimposition of new territorial boundaries on the interdependence of the physical system had obvious implications for interstate water-sharing relations, especially since these new borders allowed the Central Asian successor states to assert ownership rights over water resources within their territory and related infrastructure for the first time. The act in which Kyrgyzstan declared ownership over all the hydroelectric structures within its territory was not surprising since this is a form of claim-making and, along with the actual ability for independent policy making within territorial borders, it is an essential component of state building and sovereignty enhancement.

Yet, contrary to expectations, the removal of an external decision-making authority did not preclude cooperation by the Central Asian leaders immediately after independence. Even though Kyrgyzstan and Tajikistan as upstream riparians could have exercised their right to absolute sovereignty over their water resources, they did not behave similar to Slovakia in East-Central Europe that unilaterally diverted water from the Danube for hydroelectric production. Instead, the five ministers of water management signed the first agreement on February 18, 1992 on "Cooperation in the Management, Utilization, and Protection of Water Resources of Interstate Sources" wherein the water resources of the region were defined as "common" and "integral" (article 1). According to article 3 of the agreement, the Central Asian states "commit themselves to refrain from any activities within their respective territories which, entailing a deviation from the agreed water shares or bringing about water pollution, are likely to affect the interests of, and cause damage to the co-basin states."[46] They agreed to jointly undertake activities for the solution to the problems related to the drying up of the sea and to determine yearly sanitary water withdrawals based on the availability of water resources (article 4).

This agreement established the Interstate Water Management Coordinating Commission (IWMCC—later referred to as the Interstate Commission for Water Coordination or ICWC), which was composed of the five ministers of water management. These ministers met on a quarterly basis to define water management policy in the region and work out and approve water-consumption limits (broken down by growing and non-growing periods) for each of the republics and for the whole region (article 7).[47] The Central Asian states kept the regional centers for distribution (BVOs) as organs of the ICWC. They were largely responsible for implementing decisions regarding water sharing. The BVOs retained an executive function with regard to the operation of hydraulic works, structures, and installations on the rivers. Although with independence these hydrotechnical assets became the property of the territory upon which they were located, article 9 of the agreement transferred them to the BVOs for temporary use.[48] Between April and August 1992, additional protocols and resolutions were introduced to clarify the mechanisms for cooperation on the management, utilization, and protection of the water resources of the Aral basin and to outline joint measures for the solution of the Aral Sea problem.

It was remarkable that the Central Asian states, unlike most developing countries, rapidly concluded a water-sharing agreement. Yet this first stage of international institution building for water management had less to do with mitigating the desiccation of the Aral Sea (i.e., environmental protection) than with ensuring that cooperation would be continued for political reasons in the transitional period. The shared fears of what the future would hold in the absence of Moscow created the enabling conditions for the Central Asian states to prefer inertia to change in 1992.[49] In addition, the memory of the Osh riots coupled with the other small-scale resource conflicts loomed heavily over the leaders during the initial days of independence. The leaders sought to prevent such conflicts from taking place in the post-Soviet period. Most important, precedent or what can be considered inertia influenced the negotiations.[50] During this first phase of renegotiating institutions in Central Asia, the leaders worked within the confines of the former system, grafting new institutional structures onto previous ones in order to sustain cooperation and prevent discord. The only other system of water management that existed in the pre-Soviet period had been superseded, so the Central Asian leaders were unable to revert back to pre-Soviet practices of decentralized management.

In short, during the first year after independence rapid environmental cooperation can best be explained by inertia—not wanting to disrupt or to depart from past practices, especially since the leaders were essentially concerned with bringing in the cotton harvest. The Central Asian leaders convened in early 1992 to largely guarantee that the planting season would not be interrupted in the spring since irrigated agriculture provided the foundation of the basin's economy (World Bank 1993a, p. 19). As in the Soviet period, most resources continued to be directed toward the growing of cotton. Even though some of the raw cotton still went to Russia for final processing after the Soviet Union's collapse, the Central Asian governments actively sought alternative export routes for hard currency at world market prices. The Meredith Jones Group (UK) was one of the first foreign buyers to enter the cotton market after independence.[51] At the same time, the Israeli businessman Saul Eisenberg had negotiated deals with several Central Asian heads of state to export cotton in exchange for new irrigation technology.[52] The sale of cotton, which was firmly controlled through state marketing boards, allowed Uzbekistan and Turkmenistan, in particular, to easily substitute their reliance on Moscow with foreign buyers. In Uzbekistan, for example, all cotton sales required the approval of President Karimov and were directed through the Ministry of Foreign Economic Relations. The revenue generated from these sales helped the cotton producing states cushion the initial shock of embarking on multifaceted transitions away from state socialism.

Resistance to change was evident in an interview with Uzbekistan's Water Minister, Rim Ghiniatullin, who said: "No matter what kind of political system we will have, a centralized system for water management will still be necessary."[53] At this first stage such outlooks were widespread, since the experts were accustomed to working together in an environment relatively free of hostilities; they all knew and understood one another, since they were all trained and brought up through the same ranks of the Soviet system. Sometimes, the ties were perceived to be even deeper than just professional connections. For example, advisors to the president of Kyrgyzstan, who were demanding that Kyrgyzstan be able to sell water to Uzbekistan and Kazakhstan after independence, believed that the former Water Minister in Kyrgyzstan, Meirajdin Zulpuyev, had conceded too much to the Uzbeks in 1992.[54] They attributed this to the fact that "Zulpuyev's wife is an Uzbek" and that he came from Osh in the South along the border with Uzbekistan.

The Failure of Inertia to Bolster Regional Environmental Cooperation

From the above discussion, one can alternatively reason that large-scale conflict was unlikely right after independence because the leaders were able to tinker with the previous institutions and employ the water-allocation formula inherited from Soviet period. However, inertia cannot explain regional cooperation after 1992 because this early agreement was neither fixed nor exhaustive. In fact, it remained unclear whether cooperation would continue or whether conflict would ensue once the states began to develop their own political and economic policies associated with empirical sovereignty. As part of state making, each state began to pursue divergent strategies to accommodate the particular needs of its population. Some of the Central Asian states stressed food security whereas others gave priority to obtaining energy self-sufficiency. Already both an upstream and a downstream state wanted to expand irrigated agriculture. Turkmenistan sought to extend the Kara Kum canal in order to support the additional reclamation of land for irrigated agriculture. At the same time, the 1992 agreement notes that Tajikistan expressed its interest in increasing its water allocations for irrigated farming. Similarly, Kyrgyzstan intended to harness its hydroelectric potential. If Kyrgyzstan succeeded to unilaterally expand its hydroelectric projects, this would infringe on the downstream water users who could claim acquired rights to water allocations.

The agreement was not sustainable over the long-term, largely because of this disparity between the energy-rich states and the energy-poor states in the Aral basin. In the Syr Darya basin, the Kyrgyz government was not the only one who sought to modify both water and energy practices from the Soviet period. The energy-rich states, Uzbekistan and Kazakhstan began to formulate their individual energy development strategies, independent of each other's needs and of Kyrgyzstan's. Uzbekistan, which had to use its cotton revenue to pay Russia for oil at world prices, wanted to be self-sufficient in oil and gas production.[55] As a part of its efforts toward self-sufficiency, the Uzbekistani government created the state oil and gas company Uzbekneftegas in 1992 to increase energy production for domestic consumption. Shortly after independence, Kazakhstan pursued a completely different strategy in which it became engaged in the rapid and nearly complete privatization of its state oil and gas companies as a means to acquire foreign capital (Jones Luong and Weinthal 2001).

Even if Kyrgyzstan abided by the irrigation scheme for Toktogul and delivered water to Uzbekistan and Kazakhstan, it had no guarantee that it would receive gas and coal in return as it did under the former Soviet barter arrangements. Uzbekistan and Kazakhstan could easily cheat, since this first agreement only dealt with water and not energy allocations. Enforcement mechanisms were absent, and, for that matter, it was even unclear whether the Uzbekistan government would deliver water to Kazakhstan in accordance with the established allocations. If disputes were to arise, the agreement did not clearly lay out provisions for the settlement of conflicts of interest or for compensation for any violations. Article 13 only noted that the "managers of water management organizations" held responsibility for resolving any disputes that may arise; in practice, this refers to the water ministers (Nanni 1996).[56]

Owing to the context in which it was signed, the agreement was tenuous at best. When the water ministers originally signed the agreement, their primary concern was to ensure that the cotton harvest would be met. They did not take into account how completely integrated the water sector was with other sectors outside of agriculture such as energy. The agreement lacked legal personality, since engineers had drafted it. In addition, this first agreement was only signed at the level of the water ministers and not by the prime ministers or presidents. Customarily, the ministries and/or agencies are only responsible for implementing agreements, not signing them. The World Bank later concluded that these agreements were not adequate to constitute binding legal treaties, as they did not accord with internationally accepted standards to avoid potential conflicts in the future (World Bank 1993a, p. 5). As a result, this meant that these initial agreements left room open for further negotiation.

The newly independent governments also did not possess the domestic capacity or financial capital to enforce this agreement. The BVOs were not adequately equipped technically to carry out their roles to effectively monitoring water allocations and use among the different users, and the state environmental agencies (Goskompriroda) had lost much of their enforcement authority to monitor water quality and fine violators. Many policy makers in Kyrgyzstan also believed that the Uzbekistan government did not intend to fulfill this agreement. The director of the International Institute for Strategic Studies in Kyrgyzstan went so far as to dismiss this agreement as inconsequential and only reflective of the past Soviet practice of "friendship of the peoples."[57]

This status-quo agreement was not equitable nor environmentally sustainable over the long-term, especially if the Central Asians sought to restore the Aral Sea, at a maximum, or to preserve it at its current size, at a minimum. The immediate post-independence framework agreement incorporated the water-sharing rules applied during the Soviet period, which are based on crop requirements and quotas and pay scarce attention to water quality.[58] The Central Asian states needed to amend the agreement to ensure a minimum flow into the Aral Sea along with addressing the broader question of water quality.[59]

The Need for Aid

Because the signing of this first agreement was attributable largely to inertia, the Central Asian leaders gave preference to the retention of cotton monoculture over the basic needs of the peoples near the Aral Sea. Yet, with the cancellation of the Siberian river diversion plan and the loss of the UNEP and all-Soviet initiatives in the region, the Central Asian leaders jointly needed to take over the role that Moscow was beginning to play in bringing about a solution to the Aral Sea crisis. They lacked, however, the financial resources to carry out such a program or even to begin to undertake measures to mitigate the Aral Sea crisis as noted in article 4 of the 1992 agreement. During the Soviet period, the republican elite distributed the cotton revenue along with the transfers and subsidies received from Moscow as patronage throughout the system by either providing social services to the population or by making freely available water and energy resources and other agricultural inputs.

When the Soviet Union collapsed, the essence of this system of resource transfers was imperiled, which threatened the base of support for the new governments unless some other actor could fill the void. Indeed, the collapse of the Soviet Union suspended the Central Asian states' access to resource transfers and subsidies from Moscow. Turkmenistan, for example, was one of the main beneficiaries of transfers from the union budget in which direct transfers accounted for up to 20 percent of its total revenue or 10 percent of GDP during the end of the Soviet period (World Bank 1994b, p. 10). Compounded by the loss of resource flows, the Central Asian states experienced major disruptions in trade and payment arrangements with other former Soviet states. For example, Ukraine was a large consumer of Turkmenistan gas, but with independence it lacked foreign currency reserves to pay Turkmenistan for gas deliveries. Without a stream of these side payments from Moscow, the newly independent governments could not dole out side payments internally to the population in return for social acquiescence. The cessation in resource transfers aggravated the economic situation for the poorest of the Central Asian states—Kyrgyzstan and Tajikistan. In 1991, Kyrgyzstan's hard currency exports were the lowest of any Soviet republic ($23 million), and the lack of alternative exports such as cotton in Turkmenistan and Uzbekistan exacerbated its need for aid (Pomfret 1995, p. 60).

The Central Asian states, furthermore, began to experience similar problems to other developing countries associated with poverty, poor health care, economic collapse, and environmental degradation. Several years after independence, the United Nations Human Development Index (see table 5.3) ranked the Central Asian states as having "medium human development," with Tajikistan at the low end and Kazakhstan at the high end. After independence, Central Asia's level of human development declined steadily. Whereas during the Soviet period Moscow provided

Table 5.3
Source: United Nations Human Development Index (http://www.undp.org/hdro).

	HDI rank		Life expectancy at birth (years), 1994	Adult literacy rate, 1994	Real GDP per capita, (PPP),[a] 1994
	1996	1998			
Kazakhstan	72	93	67.5	97.5%	3284
Turkmenistan	90	103	64.7	97.7%	3469
Uzbekistan	94	104	67.5	97.2%	2438
Kyrgyzstan	99	109	67.8	97%	1930
Tajikistan	105	118	66.8	96.7%	1117

a. PPP: purchasing-power parity, in dollars.

universal education and health care, the loss of resource transfers from the center contributed to this sharp deterioration in the quality of basic health services and disruption in the educational system. Although the Turkmenistan leaders have maintained their commitment to free health care, for example, the share of its budget devoted to health care fell to 6.9 in 1992 from 11.2 in 1989 and 9.6 in 1991 (World Bank 1994b, p. 115).

All in all, independence undermined the reciprocal patron-client relationships that defined patterns of interaction among the Central Asian leaders, regional elites, and the general population during the Soviet period. Without a source of outside patronage, the newly independent leaders found it difficult to "pay off" the population and maintain the reciprocal relationship between the government and society since Moscow had been largely responsible for supplying social protection. Especially, the ruptures in trade and the loss of central budgetary transfers from Moscow in the form of patronage (side payments) and social protection weakened the system of social control in Central Asia. In response, the Central Asian leaders searched for new ways in which to deliver social services—primarily education and health care.

The Central Asian leaders, in short, needed to figure out a way to compensate those hardest hit by the Aral Sea crisis. As of July 1992, Kazakhstan officially declared the region surrounding the Aral Sea a "disaster zone." By the end of the year, the Cabinet of Ministers in Kazakhstan passed a resolution that laid out measures to provide social protection to those residing in the Near-Aral region according to whether they lived in a ecological crisis or pre-ecological crisis zone.[60] Yet in reality the government did not possess the financial means to mitigate the situation, one reason being that it also had to clean up a legacy of nuclear testing on its territory among other environmental issues. Without material or financial resources to placate those suffering near the Aral Sea, the Uzbekistan government made it incredibly difficult for foreign visitors to travel to the Aral Sea. In practice, the Aral Sea became a "closed zone" right after independence.[61] In contrast, before the breakup of the Soviet Union, the Central Asian leaders encouraged outsiders to visit the disaster zone in order to help generate international support for the Central Asians' case against Moscow for much needed assistance. For example, in

October 1990 an international symposium on the Aral Sea crisis was held in Nukus, Karakalpakstan in which foreign and local scientists participated.

The new governments sought aid at the domestic level not only to compensate those in the Aral Sea disaster zone, but to help pay off the eco-nationalists opposed to the region's reliance on cotton monoculture. Turkmenistan and Uzbekistan continued cultivating cotton because it enabled them to keep up a system of social control. Yet, by maintaining a system of cotton monoculture, the Central Asian leaders could no longer assail against what was earlier perceived as Moscow's imposed economic priorities in the region. The Uzbekistan government, for example, needed to mollify the increasingly boisterous eco-nationalist movements who were highly critical of the all-encompassing role cotton monoculture played within Central Asian society. Birlik and Erk had continued to press the government to decrease its reliance on cotton monoculture and to address the Aral Sea crisis.[62]

In summary: This initial agreement in 1992 represented a "quick response" to a very fluid and ambiguous situation. It did not constitute sustainable environmental cooperation among independent actors with well-defined interests. On the contrary, the Central Asian governments, in their unsettling physical and political situation, were increasingly discovering their own new interests. As a result, this was only an interim agreement since it possessed neither mechanisms to ensure that it was self-enforcing nor an external enforcement mechanism to coerce and impose cooperation. In the absence of either self-enforcement or a hegemon, the only other viable alternative was the international community composed of IOs, bilateral aid organizations, and NGOs.

6

The Willingness to Intervene: Paying the Costs of the Transition

Although regional cooperation took place right after independence, it entailed merely the perpetuation of past practices of water management, codified into a new agreement. Rather, because of the immediate need for aid to prevent social dislocation and economic collapse, the Central Asian leaders internationalized the Aral Sea crisis. The Central Asian states used the international community's interest in solving the Aral Sea crisis to address their own post-independence domestic problems tied to state making. This chapter explains why the international community embarked on an intricate process of building new institutions for interstate environmental cooperation. It then looks at how the international community provided aid to the Central Asian successor states—aid that was contingent on cooperation. The international community relied on a politics of inclusion in order to offset the various interests and capabilities that were emerging across states and across sectors. The Central Asian leaders accepted this aid because it enabled them to placate the short-term interests of their regional hokims and akims[1] and because it provided these actors with the means to pay off other local constituencies.

The Willingness to Intervene: The Enlarged Role of Third-Party Actors

Induced Cooperation Revisited: Multiple Actors with Multiple Interests

In the aftermath of the dissolution of the Soviet Union, multiple actors with varying interests focused their development efforts on the Aral Sea crisis, including traditional multilateral lending organizations (the World Bank and the Asian Development Bank), small Western NGOs, and multilateral organizations (the EU, the UN, NATO). The United States, Germany, Israel, the Netherlands, Japan, and Switzerland provided direct bilateral assistance. NGOs active within the basin included the Dutch organization NOVIB, the Aral Sea International Committee, ISAR, Médecins Sans Frontières (Doctors Without Borders), Mercy Corps International, Crosslinks International, and Farmer to Farmer (Winrock International).

By early 1997 the first stages of several internationally led programs were nearing completion. At the level of IOs and bilateral assistance, the World Bank had finished the preparation stage of phase 1 of its Aral Sea Basin Program (ASBP). The European Union's Technical Assistance for the Commonwealth of Independent States (EU-TACIS) Water Resources Management and Agricultural Production in the Central Asian Republics (WARMAP) project had issued a report and recommendations. USAID had begun to assert a real presence in a separate set of negotiations among Kazakhstan, Kyrgyzstan, and Uzbekistan regarding timing releases from the Toktogul Reservoir. At the micro level, the United Nations Development Program had embarked on a capacity-building program to promote sustainable development, and Western NGOs such as ISAR and NOVIB were providing numerous small grants to local NGOs.

From the outset, the West's interests in the water sector were not solely tied to humanitarian reasons. Other post-Cold War geopolitical concerns impelled many of these IOs, bilateral assistance programs, and NGOs to intervene. At a macro level, the West sought to enhance the likelihood that democracies and markets would flourish in the successor states of the Soviet Union. From the perspective of the Western democracies, they had clearly won the Cold War, and they now sought to ensure that the Central Asian states would quickly disengage from Russia's sphere of influence and integrate into an international system dominated by a liberal economic order. The West considered Central Asia a strategic buffer region, insofar as it bordered Russia, China, Iran,[2] and Afghanistan. Kazakhstan had inherited some of the Soviet Union's nuclear stockpile, and the US government was especially worried about the potential spread of nuclear weapons and weapon-grade material.[3] Finally, Kazakhstan, Turkmenistan, and Uzbekistan lie within the Caspian basin, which contains substantial oil and gas reserves, many of which were unexplored during the Soviet era.[4] In order to lessen its dependence on Persian Gulf oil, the West sought to tap into this vast and largely unknown resource.

In view of the broad array of interests in Central Asia, which was attributable to the above-mentioned geostrategic concerns, "the environment" presented an obvious opportunity for international intervention. Similar to the Soviet period, in which the environment provided a safe arena for political mobilization against Moscow, the legacy of the environment as a safe issue area quickly enabled the international community to establish ties with the Central Asian successor states. Consensus on the need to address the Aral Sea crisis was easily attained among the actors in the region and outside the region, as it was viewed as a "win-win" situation. On the one hand, the Central Asian governments desperately needed assistance to mitigate the Aral Sea crisis and to compensate those suffering most from the sea's desiccation. They also needed to revitalize their stagnating economies. On the other, Western governments and organizations sought to improve their reputation in the region and ultimately gain a stronghold in other issue areas (such as the energy sector) by concentrating first on the environment.

According to Werner Roeder, who headed the World Bank's Aral Sea Program in Toshkent, "the Aral Sea was not the worse of the problems facing the Central Asian states, but it had a name that could attract aid."[5] It symbolized the magnitude of the challenges facing the newly independent states. Inside the region, the Central Asians were also pushing for the globalization of the Aral Sea crisis. Tulepbergen Kaipbergenov wrote: "The death of the Aral threatens not only the death of the Karakalpaks and not only that of Central Asia. . . . It is already threatening global calamity."[6] As a means of bringing the Central Asians into the world community of nation-states, the Aral Sea crisis suited both local and international interests.

The crisis, moreover, exemplified how closely intertwined the environment was with issues of economic development and state security. UNEP's previous work in conjunction with the Soviet government furnished the West with a detailed picture of the Aral Sea crisis and an understanding of the limited financial resources available within the Central Asian basin states for internally addressing the environmental and economic crises at hand. According to Philip Micklin, a geographer who has worked on the Aral Sea question for several decades, "the belated Soviet effort not only provided a substantial research and data base for subsequent international and regional activities, but laid out an 'action program' that was a major help in formulating the fundamental thrust of these later programs."[7]

Owing to the early internationalization of the Aral Sea crisis, the international community knew more about the severity of the Aral Sea crisis than it knew about other environmental problems in the region and/or in the territory of the former Soviet Union as a whole (e.g., the Chernobyl disaster, which offered a mere glimpse of the problems associated with nuclear testing and the lack of proper disposal facilities for radioactive waste).[8]

Senator Al Gore had visited the region surrounding the Aral Sea in August 1990.[9] As vice president, Gore made the Aral Sea disaster zone a high priority for USAID assistance. According to US Deputy Secretary of State Strobe Talbott, it was in the interest of the United States to demonstrate good will toward and leadership in Central Asia, to develop regional cooperation among the Central Asian states aimed at preventing future conflict over water use, to deal with the largest environmental problem in the newly independent states via a multilateral effort that leverages US assistance, and to focus US assistance strategy so that Central Asians would associate US assistance with solutions to a high-priority problem.[10] By targeting public health issues such as water, the United States hoped to establish credibility in the region as an alternative partner to Russia. Paul Dreyer, then head of the Environmental Policy and Technology (EPT) Project funded by USAID, emphasized that "regional cooperation and water management are an arm of US foreign policy."[11]

Aside from garnering much needed financial assistance, the newly independent Central Asian states sought to ensure their political, economic, and ideological separation from Russia by forging ties to the various multilateral and bilateral bodies that had expressed interest in the basin's problems. International recognition of their territorial borders through admission into the United Nations was only the first step in demarcating their juridical sovereignty from Russia.[12] Likewise, joining other regional bodies, such as the Economic Cooperation Organization, increased these states' separation from Russia by symbolizing a return to the Islamic world.[13]

To further distance themselves from Russia and the legacy of the Soviet Union, the Central Asian states needed to consolidate their internal sovereignty. This required them to build the empirical components of statehood. If the Central Asians failed to secure their own borders, devise economic and social policy, and collect revenue from their populations through taxation, it would be difficult for them to break their ties with Russia, upon which they were dependent for guaranteeing their external borders and for most of their revenue during the Soviet period. Thus, by courting major multilateral organizations (the World Bank, the International Monetary Fund, the European Bank for Reconstruction and Development, the European Union, NATO, and the United Nations) to assist them with building domestic political and economic institutions, they could fortify both their territorial separation and their empirical independence from the legacy of the Soviet Union.

New states equate empirical sovereignty with the ability to make their own policies. In the water sector this translates into giving priority to independent decision making in place of joint management decisions. Now each Central Asian state could define its own strategy for water use or control over related infrastructure within inherited republican borders. Yet members of the international community feared that, as the newly independent states began to undertake national development programs in which water demands could differ from previous allocations, conflicts of interests and water disputes would arise that had not existed during the Soviet period (World Bank 1993a, p. vi). These conflicts of interest might exacerbate pre-existing conflicts, such as the resource-based conflicts in the Fergana Valley.

In response to the growing potential for conflicts of interests, the World Bank (in particular) and the international community (in general) sought to help the Central Asian states formalize environmental cooperation in the Aral basin. In its initial report, the World Bank concluded that "despite the [1992] water agreements signed after independence of the Republics, the potential for future water disputes cannot be ignored" in view of the importance of fresh water to economic development for the region (World Bank 1993a, p. ii). The World Bank anticipated conflict, and it was operating under an assumption that it needed to intervene early because national interests would develop later that could make cooperation more difficult once the previous nomenklatura was replaced.

Whereas the World Bank was the main actor in fostering environmental cooperation during the early 1990s, NGOs were also an essential element of the institution-building process insofar as such organizations interact with both IOs and states. Especially with the "greening" of the World Bank in the 1980s and the dramatic rise in environmental activism throughout the world, the World Bank could no longer ignore the impact of NGOs on environmental policy making. The World Bank must consult and include provisions for NGO involvement in many of its projects. Between 1973 and 1988, only 6 percent of World Bank projects included provisions for NGO involvement, but by fiscal year 1996 the involvement of NGOs had increased to approximately 48 percent of all World Bank projects (Prosser 2000; Reinicke 1996). The involvement of Western NGOs in the internal development of postcommunist states is notable since NGOs help both IOs and states to cope with political and economic transitions by operating at the grassroots level—a level that is often overlooked in large-scale assistance programs. These NGOs can bypass governments and politicians in order to help the populations that are at risk.

ISAR, for example, sought to support the many environmental movements to solve environmental problems locally while also helping to build a civil society.[14] Because of their earlier initiatives during glasnost, Western NGOs like ISAR and the Aral Sea International Committee could draw on pre-existing personal and organizational contacts within the basin. NGOs often conceive of their role in a different manner than the larger multilateral organizations. Their interests frequently diverge from the goals of the larger multilateral programs since they usually work with local groups that lack a "voice" in policy-making decisions. The Aral Sea International Committee saw its main role as serving as a "reminder and witness" to the World Bank's and the European Union's projects. According to its founder, it sought to ensure that the "little guys" (Karakalpakstan, the Dashhowuz region of Turkmenistan, and the Qyzlorda region of Kazakhstan) were not left out of the institution-building process.[15]

In summary: Willingness on the part of international actors to intervene existed immediately after the dissolution of the Soviet Union. The broad base of interest in helping to resolve the Aral Sea crisis for geopolitical, economic, and purely humanitarian reasons resulted in a melange of activity on the part of these multiple actors. Instead of maintaining the previous Soviet institutions for water management or even negotiating new ones among themselves, the Central Asian successor states turned to the international community to help them forge new institutions for regional cooperation and find a solution to the crisis. However, before the main actor at this first stage—the World Bank—became involved in what would be an overambitious large-scale multi-sectoral program, it insisted that the Central Asian leaders make a clear commitment to interstate cooperation. After this commitment was obtained, the World Bank made aid contingent on a firm pledge by the Central Asian states to establish new interstate institutional arrangements and attached organizations.

The World Bank and Contingent Aid

The Emergence of a New Agreement

At independence, the Central Asian leaders recognized the urgency for action to deal with the Aral Sea crisis in view of the loss of Soviet aid and the rise of environmental activism in the basin. In spite of having dealt immediately with questions of water allocation in the 1992 agreement, they had yet to directly confront the formidable challenges of reversing policies that led to the desiccation of the sea. If the Central Asian leaders continued past practices from the Soviet period, they would cooperate to destroy the environment. Having lost their funding from Moscow, they were uncertain whether they could procure water transfers from outside the basin to replenish the sea; as a result, they needed to find an alternative solution to address the desiccation of the sea.

In the wake of the agreement reached on February 18, 1992, the Central Asian governments requested assistance from the World Bank to help mitigate the ecological and health situation near the Aral Sea. The World Bank agreed to launch a mission to the region in late 1992, but it did not directly offer financial support at that time.[16] Rather, it suggested it would render assistance once the states had agreed on a new institutional framework for water management among them and had developed a list of priorities for water sharing in the region.[17] The World Bank invariably stressed the need for regional cooperation, whereas the Central Asian states only wanted aid at the domestic level (to mitigate the crisis and to replace resources from Moscow).

The World Bank could have viewed the water-sharing crisis in Central Asia only as a technical problem and a development problem. One option would have been to support the old Soviet organizational structure, with its expertise in water management, by supplying it with new equipment and financial assistance to strengthen the capacity of the ICWC and the BVOs. The World Bank, however, perceived water sharing as a political question. According to Kirmani and Le Moigne (1997, p. 14), the World Bank Mission of 1992 "stressed the need for regional cooperation and strong commitment and concerted efforts of the Republics." The option that the World Bank then followed was to make aid and its involvement contingent on the Central Asian states' devising a new institutional framework for water sharing before intervention.[18] (See figure 6.1.)

The World Bank pushed for a new agreement because the bottom line is that IOs will not give money for technical assistance in an international river basin if an agreement is not in place. The international donor and legal communities equate cooperation on the basis of fundamental principles of water law with the establishment of an international water basin institution.[19] Projects that contradict these principles of international water law are ineligible according to the operational procedures of organizations such as the World Bank. The World Bank, therefore, did not act on technical principles alone when it informed the Central Asian states that they would have to draft a basin-wide strategy before any substantive projects would be funded.

To meet the conditions for assistance, the Central Asian states supplemented the original 1992 water-sharing agreement despite the perception within the basin that it was adequate for current conditions. On March 26, 1993, in Qyzlorda, Kazakhstan, the Central Asian leaders signed an "agreement on joint activities for addressing the crisis of the Aral Sea and the zone around the sea and for improving the environment and ensuring the social and economic development of the Aral Sea region."[20]

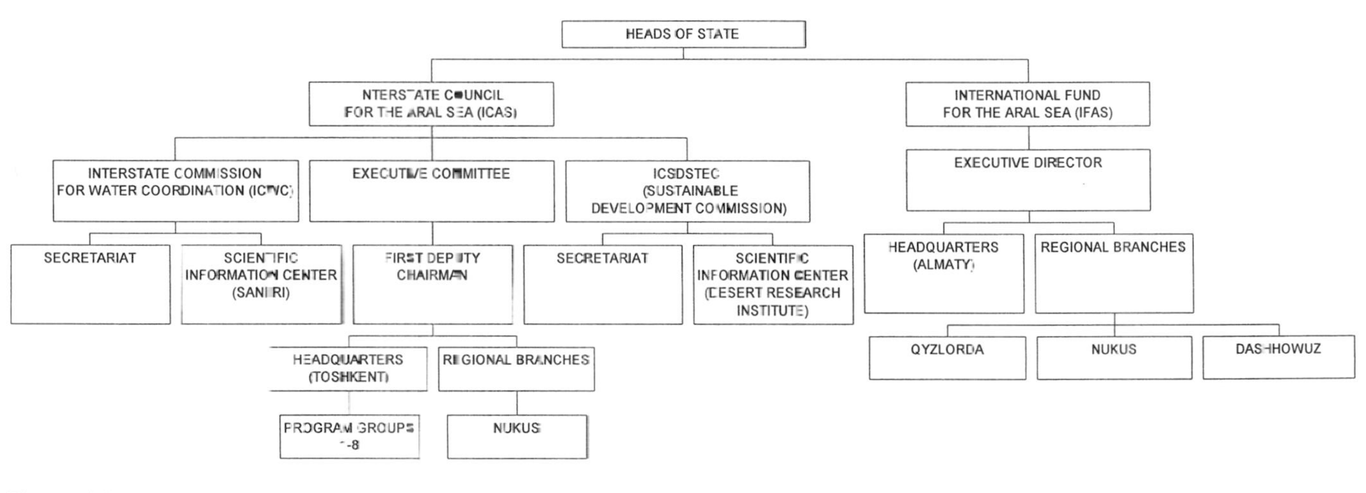

Figure 6.1
An early regional organizational chart for the Aral Sea basin (adapted from Aral Sea Program—Phase 1, Progress Report 3, February 1996).

An appeal to the international community for assistance to support this initiative followed. This agreement and the accompanying statutes created two "apex organizations" to the original ICWC: the Interstate Council for Addressing the Aral Sea Crisis (ICAS) and the International Fund for the Aral Sea (IFAS).[21] In July 1994, the Central Asian leaders established the Interstate Commission for Socio-Economic Development and Scientific, Technical, and Ecological Cooperation (ICSDSTEC), hereinafter referred to as the Sustainable Development Commission (SDC).

According to the 1993 institutional structure for water management in the Aral basin, ICAS and its executive committee (EC) became the main organizations for developing and implementing policies and programs in the Aral basin. ICAS, composed of 25 high-level representatives from the five states, was supposed to meet twice a year to discuss and decide policies, programs, and proposals put forth by the EC. The EC's charter equated it to a sovereign government with full powers to plan and implement programs approved by ICAS. The heads of the states established IFAS to finance the Aral Sea programs using contributions from the five states and from other donors. The agreement stipulated that each Aral basin state should allocate 1 percent of its gross domestic product to the fund. The Central Asian leaders nominated President Nursultan Nazarbaev of Kazakhstan as head of this fund.

The new interstate agreement signaled the Central Asian leaders' seriousness about addressing the Aral Sea crisis to the international community. Yet, because the Central Asian leaders had devised this agreement quickly, numerous internal inconsistencies remained. For example, the relationship between ICAS and ICWC was unclear in view of the duplication of functions and responsibilities in their statutes.[22] While assessing the legal basis of the agreements, the TACIS-WARMAP study characterized both this agreement and the 1992 one as framework agreements (i.e., agreements that merely establish basic principles).[23] For the international community, this meant that there was room to intervene further to clarify and strengthen the institutional arrangements for water distribution and allocation within the basin. Indeed, these new agreements prepared the foundation for initiating a subsequent process of meetings and seminars to discuss water-sharing patterns and programs for environmental mitigation in the Aral basin.

In short, the Aral basin states took these steps to demonstrate to the international community their willingness to cooperate in the management of the basin's waters and to undertake joint projects to mitigate the crisis. They designed the new organizational bodies to facilitate coordination with the donor community. Accordingly, most of the multilateral financial organizations and development agencies concentrated their efforts on bolstering these new apex organizations so that they would actually have the capacity to meet their stated objectives.

At an international seminar held in Washington on April 26, 1993, ministerial-level representatives of the various Central Asian states presented the new institutional arrangements and reconfirmed their commitment to cooperate to address the Aral Sea crisis.[24] At this point, the World Bank and a number of working groups established by the regional organizations prepared a list of specific projects for financing. In the spring of 1993, the World Bank, in conjunction with UNEP and UNDP, met with the Central Asians to devise the framework of a program for the Aral basin. The "Proposed Framework of Activities" called for seven "thematic programs" and nineteen "urgent projects."[25] The seven main thematic programs focused on the following:

developing a regional strategy for managing water resources and improving the efficiency and sustainability of dams

hydrometeorological services and regional environmental information systems

managing water quality

restoring wetlands and conducting environmental studies

clean water, sanitation, and health

integrated land and water management in the upper watersheds

automated controls of the two basin-wide agencies for water allocation (BVOs).

A supplementary program on capacity building (see appendix) was tacked on.

With the collapse of the Soviet Union and the suspension of the UNEP program, the Central Asians considered the Aral Sea Basin Program under the World Bank's auspices the preferred vehicle for finding concrete solutions to the Aral Sea crisis.

The five heads of states approved this program of concrete actions on January 11, 1994, in Nukus, Karakalpakstan.[26] The program listed four objectives:

to stabilize the environment of the Aral Sea Basin
to rehabilitate the disaster zone around the Sea
to improve the management of the international waters of the Aral basin [also referred to as strategic planning and comprehensive management of the water resources of the Amu and Syr Rivers]
to build the capacity of the regional institutions to plan and implement the above programs.[27]

The affirmation of the Central Asian leaders' commitment to work with the international community opened up the door for much-needed assistance at the domestic level. On June 23–24, 1994, the World Bank, in collaboration with UNDP and UNEP, convened a donor conference in Paris to raise the initial funds for the Aral Sea Basin Program. Implementation of the first phase of the Aral Sea Basin Program, carried out by researchers from Europe, the United States, and the Central Asian states was estimated to cost $470 million.[28]

The World Bank recognized at the outset that the Soviet legacy of water mismanagement could not be reversed overnight. Instead, it took a long-term perspective in which the program would involve three phases, the last stage continuing until the year 2025. The project included a political component. The international community could thus justify extending multilateral assistance to the region on the basis that the overall economic development of the region was inextricably tied to regional cooperation. The centerpiece of the program called for the development of a regional water strategy and a capacity-building program for these apex organizations and regional distribution centers (i.e., the BVOs).[29] In short, the World Bank described its Aral Sea Basin Program as "designed precisely to promote this regional cooperation."[30]

Unlike the February 1992 agreement, this 1993 agreement and the subsequent ratification of the Aral Sea Basin Program were signed at the level of heads of state. Yet this 1993 agreement would not have been signed without the prodding of the World Bank and the lure of potential funds. Because water is such an important and sensitive resource for the economic development of the region, outside intervention began at this level.

Kirmani and Le Moigne (1997, p. 16) described early World Bank missions to the region as a form of "quiet diplomacy" in which the World Bank's Director and Vice President conducted negotiations with the five heads of states. The World Bank realized quickly that, owing to the inherited legacy of top-down decision making, only the Central Asian leaders had the authority to negotiate agreements and to formulate policy. Without approval from the highest echelons in Central Asia, lower-level bureaucrats would not freely discuss or share information and data. The World Bank's immediate intervention and its willingness to provide aid contingent on a real commitment for regional cooperation at the highest levels led to the emergence of a new framework for water sharing in the Aral basin.

An Alternative to Inertia: Shifting the Feasible Set of Solutions

In view of the World Bank's early perception of the situation and of the comments of some who sat in on the meetings, the following counterfactual scenario is plausible: If the international community had not taken on such a large presence, the Central Asians might not have perceived the immediate need to exhibit overt cooperation. Instead, the newly independent Central Asian states might have shown inertia, discord, or non-institutionalization.

Indeed, the World Bank (1993a, p. 13) noted that the Central Asian states had presented "a united front" to the international community in spite of some underlying differences among them concerning future patterns of water use and allocation. Although all the riparian states embraced the rhetoric of saving the Aral Sea, they were not equally committed to actually restoring it to its pre-disaster conditions (ibid.). Nonetheless, the show of unity continued during the first phase of the World Bank's Aral Sea Basin Program. In an interview with a local World Bank consultant, Elmira Nouritova of the World Bank remarked that, even though the participants in the various working groups might have been fighting among themselves, when dealing with the international participants they reconciled their differences.[31]

Similarly, the World Bank observed during the initial negotiations over the scope of the program that the Central Asians often overemphasized the severity of the crisis as a means of attracting assistance. Without a doubt, the UNEP diagnostic study accurately depicted what had happened in the Aral Sea as one of the worst environmental disasters ever. But rather than focus on the domestic roots of the problem associated with cotton monoculture and state controlled planning, the Central Asians continued to hyperbolize the situation as an international calamity. Both government officials and local activists perceived that solutions to the Aral Sea crisis were beyond the scope of local efforts. When I was touring the disaster zone in 1994, local members of the population consistently reinforced the prevalent perception that nothing could be done to save the Aral without outside assistance. In Nukus, a member of the Union for the Defense of the Aral Sea and Amu Darya summed up the sentiment of the group this way: "We do not need further research, but instead we need action—technology and money—the basic things to solve the problem. It is more important to shape public opinion in other states like the United States in order to get help rather than to focus on the government in Uzbekistan."

Whereas during the Soviet period the Central Asians rallied around the Aral Sea to demand redress from Moscow, after independence they used the Aral Sea disaster as the most obvious vehicle for garnering international aid. The World Bank accordingly recognized that, if the Central Asian states actually had to pay the costs of providing a solution, their unified approach could prove "fragile" (World Bank 1993a, p. 14). In summary: The newly independent states were betting that the international community would provide a solution to the Aral Sea crisis, so that they would not have to devise one on their own.

In post-Cold War environmental disputes, in contrast with many other types of political or economic disputes, international organizations have provided more than financial and technical assistance. In the Aral basin, the World Bank's perception of the situation as acute led it to deviate from its often-cautious approach to international intervention. Since its "greening" in the mid to late 1980s, the World Bank has shifted its focus from funding purely "development" projects to funding projects that have a large environmental component. Since the 1992 United Nations Conference on Environment and Development in Rio de Janeiro and the subsequent creation of the Global Environmental Facility as a funding source for environmental projects, the World Bank's role as a proponent of environmental projects has increased. The timing of the collapse of the Soviet Union and the subsequent Rio conference created an extraordinary moment for the World Bank to take on a massive environmental project that could link together economic development, conflict prevention, and environmental protection.[32] Since the international donor community was already focusing on aiding these new economies in transition, environmental protection became one part of the donor assistance programs to Eastern European and the Soviet successor states that would enable them to advance broader economic and political reform.

The Aral Sea crisis, moreover, provided an unique "test case" to link economic and political reforms with environment and conflict issues and thus enabled the World Bank in 1992 to adopt a very uncharacteristic role in Central Asia. According to Kirmani and Le Moigne (1997, p. 15), the World Bank was an active participant in assisting the negotiations over proposals to address the Aral Sea crisis: rather than "act as a passive neutral third party," it "played a proactive role in search of development alternatives acceptable to the parties." In short, the World Bank helped to shift the feasible set of solutions to the problem of water management in Central Asia rather than allowing the Central Asians to retain past water-management strategies because of inertia.

Specifically, the World Bank was not willing to fund the Central Asians' two main proposed solutions to "save the Aral." The Central Asians were hoping either to raise funds to reinstate the Siberian rivers diversion project or to pump water from the Caspian Sea to the Aral Sea. During the early mission to Central Asia in 1992, the World Bank encountered much reluctance among the Central Asians to removing the option of outside water transfers from the agenda. Many scientists and many bureaucrats assumed that the international community would replace Moscow as a dispenser of financial resources to enable them to continue working on schemes to bring water from outside the region to the Aral Sea.[33] Here, the Soviet legacy of relying on a "technical fix" to transform and control nature persisted, and it shaped perceptions of nature within Central Asian society. For example, the director of the Institute for Water Problems in Toshkent, Najim Rakhimovich Khamrayev, blamed Russia for the water crisis in Central Asia, even after independence. He asserted that Russia had failed to deliver the Siberian rivers to

Central Asia.[34] As an alternative to this water transfer within the former Soviet bloc, Khamrayev proposed an inter-basin transfer of water from the Indus River to the Aral Sea. When further pressed about the problem of political boundaries as an impediment to inter-basin transfers, he dismissed this issue as inconsequential. Such technical solutions did not reside only within the scientific institutes; they also percolated into the policy realm. Representatives of the Institute of Strategic Studies—at the time, the preeminent institution advising the president of Uzbekistan—mentioned the possibility of inter-basin transfers from the Siberian rivers or from the Caspian Sea as a means of addressing the Aral Sea crisis.[35]

By shaping the form and the scope of the solution to the Aral Sea crisis, the World Bank acted as the driving force behind the cooperation in which the Central Asians were engaged. The World Bank argued that the Aral Sea could not be saved and instead pushed for a program for regional cooperation to mitigate the effects (Kirmani and Le Moigne 1997, p. 15). Moreover, the World Bank's alternative solution entailed lessening agricultural dependence on cotton while introducing more efficient water management techniques through infrastructure improvements and provisions of incentives for conservation. According to the World Bank's proposal, the water saved by shifting agriculture away from growing water-intensive crops and increasing the efficiency of gravity irrigation would help to stabilize the sea, but would not restore it to its pre-1960 level.

From the perspective of the Central Asian leaders this was the least favored solution, since restructuring agriculture could potentially undermine the system of social control. Reforming the Soviet command economy required rethinking past practices and incorporating economic and social costs into policy decision making—two areas that the Central Asian leaders viewed as undesirable after independence. From the perspective of the environmental groups and eco-nationalist movements, this also was not their most preferred solution. Although they advocated a reduction in the amount of acreage designated for cotton cultivation, their platform was premised on saving the Aral Sea. Even in 1995, Pirmat Shermukhamedov was advocating a canal to transfer water from the Caspian as the appropriate solution to revive the Aral.[36]

In addition to inducing the states to constitute a new interstate agreement, the lure of side payments in the form of material and technical assistance helped shift the priorities for the types of proposed solutions to mitigate the Aral Sea crisis. In order to achieve international involvement in the Aral Sea crisis, the Central Asian leaders and policy makers accepted the fact that the Aral Sea would not be saved by inter-basin transfers. At the same time, they needed to present a unified approach, which required adopting the proposed framework for addressing the Aral Sea crisis. For many Central Asians, international involvement conjured up expectations of an overnight solution to an entrenched problem caused by more than 40 years of mismanagement and an unrelenting faith in the power of technology. Thus, by restricting what the World Bank was willing to finance, it and other international actors influenced the scope and form of possible solutions.

Other Donors, Further Intervention

The World Bank's commitment to the Aral Sea Basin Program brought with it other international actors willing to intervene. As result of the Aral Sea Basin Program, the European Union's Technical Assistance for the Commonwealth of Independent States (EU-TACIS) initiated its WARMAP project in 1994 to support the EC of ICAS and cooperating institutional bodies (at both the regional and national levels).[37] The TACIS-WARMAP project focused its attention on capacity building, the development of strategies for managing water resources, the creation of a management information system for the EC-ICAS, and the improvement of water efficiency in the agricultural sector.[38]

Unlike many of the other aid projects with a purely technical focus, TACIS-WARMAP had a specific legal and institutional agenda to devise a framework for water sharing based on legal principles in accordance with the 1966 Helsinki Rules and the International Law Commission recommendations.[39] It set out to codify water allocations so that there would be a permanent mechanism for water divisions (quantifiable minimum releases of water into the Aral Sea) even if national interests changed over time.[40] Traditionally, each year the ICWC had renegotiated water allocations, including how much river flow will ultimately reach the sea. TACIS-WARMAP also sought to ensure that national water laws conformed to the international agreement since variation existed at the national levels. After independence, Turkmenistan was the only Central Asian country to maintain the 1972 Soviet Water Code while the others adopted new water codes or laws as part of the process of constituting new domestic institutions for sovereignty enhancement.[41] If the new domestic laws were not harmonized across borders, this would preclude implementation of the new interstate agreements. Finally, TACIS-WARMAP sought to deal with the issue of overlapping jurisdiction among the institutions and organizations regulating the Aral basin.[42] These contradictions were largely the result of the Central Asians negotiating these agreements to serve political reasons and not to conform to principles of international water law.

Working in conjunction with the World Bank, the UNDP initiated an "Aral Basin Capacity Development" project.[43] It sought to strengthen the capacity of the Sustainable Development Commission (formerly ICSDSTEC) and the ICAS. Its main focus centered on the promotion of sustainable development, a concept that only began to receive attention with the entrance of the international community in Central Asia. In 1995, the UNDP organized a conference on sustainable development in which all the presidents signed the Nukus Declaration, renewing their overall commitment to the Aral Sea Basin Program.[44] According to article 1.8 of the declaration, "the Central Asian states recognize the previously signed and operation agreements, treaties, and other legal documents which regulate their relations in the sphere of water resources in the Aral basin and accept them for implementation."[45] In addition, they stressed "a need for an international convention on the sustainable development of the Aral basin" (article 1.10). Momir Vranes, a UNDP Program Manager in Environment and Resources Management, described such declarations as an "exercise in conflict prevention."[46] Here too, the international community pushed for a symbolic overture at the regional level as a precursor to the commencement of an internationally sponsored development program.

Although the Central Asian states quickly coordinated water policy strategies after the breakup of the Soviet Union, both the Central Asian leaders and the international community recognized that the continuance of past policies would impede economic development and could potentially lead to new disputes over water allocations. USAID, in response, chose to focus specifically on this question and to use its strategic resources to encourage local actors to consider new options for water sharing and to demonstrate the linkages between the energy and water sectors. It concentrated its efforts on the potential and real disputes over water management schemes for the Toktogul Reservoir. Rather than assuming that the 1992 water-sharing agreement was fixed and exhaustive, USAID acted as a catalyst for a new set of negotiations that are discussed in detail in chapter 7. To set the ground for these negotiations, USAID carried out a series of workshops on water policy and pricing issues.

Overall, the international community leveraged its assistance to foster regional cooperation and institution building for addressing the Aral Sea crisis and to modify Soviet approaches to problem-solving. Negotiations over water institutions in Central Asia have not been only between states. Rather, multiple actors have abetted the institution-building process, and in the post-Cold War period multilateral approaches have complemented and in some instances replaced traditional bilateral forms of diplomacy (Weinthal 2000).[47] The Aral basin case bears out the premise that the negotiating arena over environmental resource issues such as water institutions is much broader than traditional security or economic situations, especially since environmental negotiations embrace multiple actors ranging from multilateral lending organizations to NGOs.

Mechanisms of Intervention at the Interstate Level: Balancing Competing Interests and Capabilities

Policy of Inclusion

Third parties alter bargaining situations over shared resource systems by offering selective incentives and resources to the various parties that can potentially undermine an agreement. This is where the technical and political aspects of water management mesh. Developing countries require technical capabilities and the financial resources to carry out water management schemes, and IOs, bilateral aid organizations, and NGOs provide the technical know-how and the foreign capital to facilitate these water development programs. In Central Asia, such third-party actors were able to participate in the water-sharing process because they used the lure of financial and technical resources to balance newfound asymmetries of interests and capabilities.

Overall, different actors pursued different forms of intervention within the basin. The World Bank and TACIS established relatively broad multilateral projects that focused on technical assistance and institution building. There were two aspects to the World Bank's support. First, it promoted regional programs through the Global Environmental Facility. Second, these programs were smaller in comparison to the World Bank's investment lending for projects in water supply, irrigation and drainage, and other projects at the national level on human needs and poverty alleviation.[48] USAID also concentrated its efforts on two levels. First, it undertook bilateral, smaller-scale projects that were much more visible, and second, it promoted long-term cooperation over water releases from Toktogul. UNDP along with other foreign NGOs converged on the creation of local NGOs and community participation in addressing the Aral Sea crisis.

Donor support pledged at the Paris conference in 1994 for regional level activities amounted to $32 million. During the first phase of the Aral Sea Basin Program, USAID contributed $7 million for water supply, energy and water management policy, and health programs; the Netherlands pledged $6 million for water quality assessments, preparation of wetland restoration, capacity building, and UNDP support; EU-TACIS granted $7 million for its WARMAP project focusing on preparation of interstate agreements, regional water and land database, pilot projects, and monitoring of on-farm water management issues; the World Bank donated $5 million which was a special grant for institution building; UNDP offered $2 million; and a series of other donors with assistance totaling about $6 million (Canada, Finland, Switzerland, UK, Italy, Denmark, Sweden, Japanese PHRD funds, and the Kuwait Fund) chose to support regional activities (World Bank 1998, p. 1).

All the internationally led programs had to balance a multitude of domestic interests who eagerly wanted a piece of the foreign aid pie. As a result, IOs ensured that each state received its fair share of the donors' package, which at times entailed trying to uphold the ethnic, regional, and national balance among the Central Asian states. This policy of inclusion and fairness premised on ethnic, regional, or national differences was rooted in the Soviet legacy in which Soviet policy toward the republics granted specific privileges and recognition to each ethnic or national group as a means to bolster their ethnic or national culture. For the water sector, this meant that no state could be given preference over another, and as a result, each downstream state received one of the new interstate organizations.

At the interstate level, the newly created apex organizations encapsulated the power asymmetries between the countries and the variation in ethnic and national cleavages within the organizations. By constituting three new organizations instead of one overarching body, this guaranteed the downstream states equal access to international funds. In many ways, this mirrored the Soviet process of reproducing parallel administrative bodies across republics. In the case of these new apex organizations, the Central Asian leaders placed ICAS in Uzbekistan, IFAS in Kazakhstan, and the SDC in Turkmenistan. The World Bank largely focused its activities on both ICAS and IFAS whereas TACIS-WARMAP concentrated its efforts on ICAS. UNDP targeted the SDC in order to foster sustainable development while also working with ICAS.

The Central Asians, moreover, constructed these organizations and signed new agreements in response to what they perceived as the role of states in the international system of nation-states. They constituted new organizations as a sign of becoming "legitimate" and "recognized" nation-states rather than appendages of Russia. Yet, because these organizations were fabricated to meet the conditions for international aid and intervention on the part of IOs, many of them remained dormant during the first few years after independence. As of March 1995, Bulat B. Turemuratov of IFAS pointed out, IFAS had still not collected the money owed to it by all the states to sponsor projects to improve the livelihood of the populations living near the Aral Sea.[49] Turkmenistan refused to send its money to IFAS, and said that it would use its designated funds for projects exclusively within Turkmenistan. Tajikistan was unable to meet its commitment as its economy was completely ravaged by the civil war taking place within its political borders. Without being able to carry out its function, IFAS was largely dependent on assistance from the World Bank that, for all intents and purpose, managed the fund. A Kazakhstani geographer who worked extensively on the Aral Sea described IFAS as "an additional structure created to give something to Kazakhstan in order to prevent confrontation among the republics. It is 2 years old, but basically, they do nothing because they have no money. Yet, they have an office and cars."[50]

Besides supporting the three new organizational structures, IOs had to be sensitive to staffing issues. During the Soviet period, most of the main scientific research and training institutes were located in the downstream states. For example, the all-Central Asian Institute for Irrigation Research (SANIIRI) and the Toshkent Institute of Engineers of Irrigation and Agricultural Mechanization were based in Uzbekistan while the Institute for Desert Studies was located in Turkmenistan.[51] Since institutes like SANIIRI generated most of the information and data relating to the water system and irrigation use, IOs providing technical and financial assistance designated Uzbekistan as their base for operations. Although SANIIRI was an all-Soviet institution before the collapse of the Soviet Union, it status shifted to Uzbekistan after independence. This enabled some of the Uzbekistani organizations to assume a greater role in the negotiating process and to ensure that their vested interests were clearly represented. For example, Viktor Dukhovny, the head of SANIIRI, was appointed the leader of the World Bank's working group on planning a regional strategy (Working Group 1). At times, this produced much frustration among the other four Central Asian states and even among environmental organizations and groups within Uzbekistan that found it objectionable that the "old guard" continued to exert its influence on the system of water management in the post-Soviet period. Dr. Vladimir Gregorovich Konyukhov, vice-chairman of Goskompriroda in Uzbekistan, claimed that these water specialists from the Soviet period only adopted the jargon and "donned the clothes of ecology" in order to ensure participation in the World Bank's program.[52]

In order to compensate for and offset the predominance of Uzbekistani specialists in the water sector, the composition of the other new organizations was also based on national and/or ethnic criteria. According to Turemuratov from IFAS, it was a "political decision" to ensure that everything was balanced among the former republics.[53] As Kazakhstan received IFAS, its staff consisted solely of Kazakhstani nationals. Uzbekistan received ICAS, and the staff of the EC-ICAS was Uzbekistani citizens and based in Toshkent. Yet, in order to keep Turkmenistan involved, its water minister, A. Ilamanov, was appointed the chairperson of the EC-ICAS even though he remained in Ashgabat, Turkmenistan. Moreover, the SDC was based in Turkmenistan. Although the donors were willing to support all three organizations, some foreign advisors like Professor Dante Caponera, an internationally renowned water lawyer, noted that in order to improve their overall capacity and turn them into international bodies with legal personality, the staffs of the respective organizations should not be made up of only one nationality (Caponera 1995, p. 28).

Finally, many of the major research institutions were and remained dominated by Russians, and this elicited a fair amount of resentment from the titular nationalities. During one informal meeting, a scientist from SANIIRI described a situation in which a "secret letter" was sent to the head of Minvodkhoz stating that Uzbekistan's image was tarnished by having ethnic Russians as the head of both SANIIRI and Glavgidromet. Indeed, many of the scientists at SANIIRI and Glavgidromet were ethnic Russians, but shortly after independence, a process of Uzbekification was slowly taking place within the ministries and scientific institutions as part of the state formation process in which to counter Russian dominance. For example, the Uzbekistan government sought to revitalize other water institutes like the Institute for Water Problems of the Academy of Sciences by staffing it primarily with ethnic Uzbeks.[54] In short, the international community had to be particularly conscious to the fact that the upper echelons of the Central Asian water sector were largely Uzbekistani and Russian.

Concerning the specific programs, the World Bank made it clear from the outset that it would not finance large-scale engineering projects. Rather it emphasized the creation of teams of local experts to identify and prepare specific projects, especially to devise a regional strategy for water management in the Aral basin. Working Group 1, for instance, received the crucial task of devising a regional strategy for cooperation in the Aral basin—to coordinate water policies among the Central Asian states.[55]

Much of the outside intervention during this first stage centered on the organization of working groups, seminars, and training sessions. Experts from the different ministries, research institutes, and sectors of the economy were all posturing to be included in the various working groups. Again this meant that the World Bank and TACIS had to balance all the competing interests when deciding who should or should not be included in decision making. The World Bank working groups, as a result, involved a large number of participants equally represented from the basin countries. The working group for the "strategy of rational utilization and protection of water resources of the Aral basin" consisted of 20 local participants in which four people represented each country, usually from the various water ministries and design institutes.[56] Similarly, Program Three on "management and estimation of water quality" included three representatives from each country, largely hydrotechnicians.[57] One World Bank report quoted that over 500 local experts in over 100 institutions were involved in all five countries (World Bank 1995b, p. 16). TACIS emphasized that just in Project One of its WARMAP program (regional strategy), it contacted 26 institutions in the five states and fifteen of these have provided over 160 local experts.[58]

Most of the participants came from the main Soviet water institutions. Yet these institutes were hit the hardest by the loss of funding from Moscow. Many of their scientists and staff left owing to declining salaries and better opportunities in the commercial sector. Kazgiprovodkhoz (the main design institute in Kazakhstan) during the Soviet period employed 1300 people, but after independence only 150 people continued to work there. According to a chief engineer, they used to build dams and reservoirs like the Corps of Army Engineers in the United States, but with independence, they have basically turned into a consulting firm in order to find replacements for the loss of Soviet investments.[59] For those that remained, inclusion in the working groups and seminars sponsored by the international community provided them with much needed additional income and employment.

For international actors to make headway and establish fruitful working relationships with these newly independent states, this approach of inclusion proved essential. USAID faced similar constraints in organizing its endeavors in Central Asia. It also needed to ensure that all the countries were equally represented within its programs, and thus it deliberately rotated its workshops among the capital cities to secure participation.

After the first workshop on information management issues was held in Uzbekistan in December 1994, USAID placed the next one in Turkmenistan in May 1995 so that the Turkmenistanis would definitely attend.[60] From the local perspective, the location in which a meeting was held carried much symbolic weight. The third workshop on pricing issues then occurred in Kyrgyzstan in November 1995. USAID has generally considered these meetings to be instrumental in paving the way for later negotiations over a new water-release scheme from Toktogul.[61]

Sponsoring Development Projects

Besides a policy of inclusion, another way in which international actors facilitate cooperation and offset various asymmetries of power is through sponsoring development projects. This is especially important for inducing upstream states, which might have to relinquish their natural upstream advantage, to participate in the institution-building process. If Kyrgyzstan, for example, were to build new dams to generate hydroelectric power or even continue to run Toktogul in the winter, this would restrict water flows downstream. By choosing whether or not to fund such projects, the international community can directly affect water-sharing patterns. International law, which prescribes the need for agreement from all affected parties before such a development project can take place, offers a lever for IOs in which to easily refuse the disbursement of funds.

By refusing to support these upstream projects, the international community, however, needed to compensate Kyrgyzstan in other ways to gain its participation. Since the Central Asian states are poor, the international community offered side payments to Kyrgyzstan to prevent it from exercising absolute sovereignty over the upper reaches of the watershed. One particular way in which the World Bank compensated the upstream riparians, Kyrgyzstan and Tajikistan, was to include provisions in the Aral Sea Basin Program for specific projects in the upper reaches of the basin. Program Six focused solely on integrated land and water management in the upper watersheds and Program One included provisions for dam safety.[62] Otherwise, Kyrgyzstan and Tajikistan lacked incentives to participate in the water negotiations for they are so far physically removed from the immediate effects of the Aral Sea crisis.

If an upstream riparian is less powerful militarily and economically than a downstream riparian, IOs, furthermore, can facilitate new negotiations over water-sharing institutions. Here, USAID came to the defense of Kyrgyzstan in its sponsored negotiations over water releases from Toktogul. Parallel to the World Bank's program, USAID initiated a whole new track of negotiations under the auspices of the Interstate Council for Kazakhstan, Kyrgyzstan, and Uzbekistan (ICKKU). By focusing on a burning issue for Kyrgyzstan—the use of its hydroelectric resources, USAID elevated Kyrgyzstan's interests on the interstate policy agenda.

Welcoming Intervention at the Domestic Level

International Intervention and State Sovereignty

Starting from the premise that the Central Asian states were weak states in need of financial resources to cushion the effects of disengagement from Moscow, my approach to two-level institution building posits that the Central Asian states would welcome international intervention rather than resisting it because of sovereignty concerns. At the most critical level, the Central Asian leaders needed to placate the hokims and akims that served as the link between the leaders in the capital cities and the general population. As a result, the leaders rushed to find new markets for their cotton and haphazardly negotiated cotton deals with several foreign buyers. Both President Karimov and President Nazarbaev signed contracts with the Eisenberg Company (Israel) in which cotton would be exchanged for new irrigation equipment and tractors.[63] Regarding the agreement between the government of Uzbekistan and Eisenberg, the hokim from Andijon Oblast was also included in negotiating this deal while the Ministry of Agriculture was excluded.[64] Such deals provided an immediate source of revenue for the designated hokims and a means to keep the people on the farms in some of the most densely populated areas within Central Asia: the Fergana Valley in Uzbekistan and Shymkent Oblast in Kazakhstan. Turkmenistan also aggressively sought foreign markets for its cotton harvest. Right after independence, it bartered most of its cotton exports with Italy, Argentina, and Turkey in exchange for plant and equipment for processing cotton; only four percent of the processed cotton produced ended up in domestic markets (IMF 1994, pp. 7, 35).

At the same time that the Central Asian leaders were forced to take over the role that Moscow played as the dispenser of social protection and patronage, the leaders were unable to blame Moscow for the multifarious mix of social and economic problems gripping Central Asian society. Instead, the Central Asian leaders needed to demonstrate to their populations that they themselves could tackle the domestic health and environmental problems precipitated by the Aral Sea crisis. This dire need to replace the loss of patronage from Moscow and simultaneously find a solution to the Aral Sea crisis provided the impetus for the Central Asian states to invite in the international community and to sign new agreements and declarations as symbols of action. The Central Asian states, thus, willingly agreed to formulate new interstate agreements because they knew that in return, they would receive financial assistance in the domestic realm, especially for the regions hardest hit by the Aral Sea crisis. With a pledge from the international community to help mitigate the crisis, the leaders could claim to their populations that they themselves were working to ameliorate the effects of the Aral Sea crisis while consolidating independence from Moscow; at the same time, they also continued to internally extract cotton rents.

While having to demonstrate to their domestic constituencies a concrete action plan for addressing the Aral Sea crisis, the newly independent states also needed to initiate concomitant political and economic transitions as part of the empirical component of the state-building process. Another reason then for inviting in the international community was the urgency for international assistance more broadly. International intervention enabled these states to embark on domestic transitions of disengagement from the legacy of Soviet rule. IOs, bilateral assistance programs, and Western NGOs helped the Central Asian states to transform domestic institutions and reformulate domestic policy in the water, agricultural, environmental, and energy sectors. For example, at the same time, that the EU TACIS had embarked on its WARMAP program, it also had simultaneous projects on energy, agriculture, social welfare, civil service reform, privatization, and banking reform in several of the Central Asian republics. Foreign advisors and consultants were working closely with the ruling elite to constitute new domestic institutions and policies in many of these sectors.[65]

As previously mentioned, in international politics those who gain by cooperation must devise incentives to make those who lose play the game. Again the process of state building is fundamental for understanding how third-party actors generated interstate cooperation through the use of financial and material resources at their disposal. In Central Asia those who gained by receiving more water are the downstream states, but in order to receive more water, they needed to induce the upstream riparians to cooperate for joint management of the basin. However, the downstream states themselves were also weak and poor and were reluctant to compensate Kyrgyzstan for water releases during the summer. Instead, the international community facilitated cooperation and strengthened state sovereignty in Central Asia through the provision of side payments to Kyrgyzstan and then later through a strategy of issue linkages. In chapter 7 I show how USAID encouraged new negotiations over Toktogul through issue linkages, which helped to induce cooperation among Kyrgyzstan, Kazakhstan, and Uzbekistan along a separate track where new asymmetries of interests and capabilities might have led to discord.

Yet the mere existence of third parties or available aid will not always lead to cooperation. In fact, these same actors were unable to achieve interstate cooperation in the Fergana Valley as a confined and bounded entity. After the initial success of the Aral Sea Basin Program, UNDP established a regional cooperation program in the Fergana Valley among Kyrgyzstan, Tajikistan, and Uzbekistan. However, immediate cooperation did not transpire. In spite of the lure of further development assistance, Uzbekistan refused to participate. Why could third parties induce rapid regional cooperation over the Aral basin and not in the Fergana Valley? In short, several factors that are related to facilitating conditions explain the difference in outcomes. First, the Aral Sea afforded a visible "crisis" for internal and external interests to converge. Besides the ethnic riots during the glasnost period, the Fergana Valley lacked a "smoking gun" that needed to be "fixed." Second and more important, how the underlying capabilities and interests line up differ in the two cases. In the Aral basin, the asymmetries of capabilities and interests between the upstream and downstream states were offsetting. Uzbekistan, the more powerful military and economic state, found itself in a weaker position in relation to the physical situation of the Aral basin, which allowed for external actors to offer side payments to equalize the bargaining situation. These offsetting asymmetries were absent in the Fergana Valley. Instead, Uzbekistan is clearly the dominant player in the Fergana Valley owing to its military and economic prowess along with having a large concentration of ethnic Uzbeks there; approximately, three-fourths of the population is ethnic Uzbek. Third, Uzbekistan fears any external involvement in the Fergana Valley that could break down its system of social control. Here, most of the Uzbekistani population is employed in the agricultural sector and in the production of cotton, which also brings in much needed foreign revenue.[66] Uzbekistan has therefore willingly refused aid because the need to protect its mechanisms for social control and for procuring foreign revenue outweighs development assistance. In contrast, assistance channeled to those hardest hit by the Aral Sea crisis has reinforced rather than undermined Uzbekistan's hold on social control. Finally, because Kyrgyzstan and Tajikistan were in dire straits for aid, when offered a chance to garner new development assistance, they had no alternative but to sign onto any form of a new international aid program even if it might compromise domestic sovereignty. Since independence, the international community has awarded Kyrgyzstan with several assistance packages to undertake economic reforms. Within days of the introduction of its new currency on May 10, 1993, the IMF approved a $23 million loan and a $39 million stand-by credit, and the World Bank announced a $60 million credit.

Indeed, in situations in which an upstream state desperately requires aid, it will be more willing to coordinate its policy and relinquish absolute sovereignty. IOs, accordingly, played a significant role for countries like Kyrgyzstan that would barely exist without aid.[67] Besides including an upper watershed project in the Aral Sea Basin Program, international actors assumed a large role in helping to restructure domestic institutions indirectly tied to the water sector. For example, in order to help Kyrgyzstan carry out its early reforms to privatize its state and collective farms, the EU-TACIS rendered assistance to restructure entirely the Ministry of Agriculture.[68] TACIS also carried out several livestock projects in Kyrgyzstan, as this was the main bulk of Kyrgyzstan's agricultural base.[69] During the Soviet period Kyrgyzstan specialized in the production of meat and wool; with 63 percent of its population rural, assistance in the agricultural sector was necessary to impede the deteriorating economic conditions facing the country with the loss of subsidies from Moscow.[70] Although Kyrgyzstan is an upstream riparian seeking to harness it water resources for hydroelectric energy, it, nevertheless, needed to be sensitive to pressure from the international community precisely because of its dependence on foreign aid.

In summary: Rather than resisting intervention in the domestic sphere, the newly independent states encouraged international actors to take on an enlarged role in helping the Central Asian states redesign some of the empirical elements of statehood. Unlike other periods of state making, the construction of nation-states in post-Soviet Central Asia is not only about rulers bargaining with their subjects; rather, both the juridical and empirical components of state building are embedded within the international context of the world polity of nation-states. The process of building and fortifying the empirical components of state sovereignty in the post-Cold War period involves an active role for IOs, bilateral aid organizations, and NGOs. They help to build and construct states through the transference of knowledge and financial and material assistance at the domestic level through their development programs.

Side Payments at the Domestic Level

Besides welcoming international assistance to fill the void left behind by Moscow and to reinforce state sovereignty, my approach to two-level institution building expected that IOs and NGOs will channel side payments to the domestic level. Side payments form the crucial link between the international and the domestic level since the Central Asian leaders used them to compensate those undermined by the transition. In Central Asia, IOs such as the World Bank or bilateral assistance programs like USAID needed to fund and target a broad spectrum of domestic actors to induce the Central Asian leaders to reach and support an international agreement governing water use. They have included the old water nomenklatura in the negotiations along with the regions hit hardest by the Aral Sea crisis. In addition to IOs and bilateral aid organizations, Western NGOs occupied a critical niche in the institution-building process, especially where they assisted those suffering in the near-Aral region. Furthermore, they supported various actors who were deliberately excluded from the bargaining process over institutions for water cooperation. In short, they served as a link between the most marginalized actors in the Aral basin and the international community that controlled the purse.

Without internal financial resources at their disposal, the Central Asian governments turned to the international community to placate the multitude of domestic interests in the water and related sectors. As described above, IOs provided side payments to a wide array of domestic actors with vested interests in the water sector through inclusion in the working groups and projects of the Aral Sea Basin Program. If the international community had ostracized the old water nomenklatura as many Western and Central Asian environmentalists favored, this would have shattered any chance of implementing a multilateral and multi-sectoral program to ameliorate the Aral Sea crisis. Even though state breakup unsettled the political and physical borders, it did not dismantle the previous system of internal governance and the patterns of water sharing by highly integrated users in Central Asia. When the collapse of the Soviet Union left the system of water management in place without an overarching authority to guide it, the Central Asians sought to find a new outside patron to finance its operations rather than to restructure the Soviet water institutions that supported a system of cotton monoculture.

Thus, in seeking to prevent the outbreak of conflicts over water, the international community did not alienate those with vested interests in the water sector, and the international community did not challenge the creation of these new apex organizations, albeit recognizing their limitations and internal contradictions. Rather, the international community continued to support the old guard and prop them up by allowing for new structures to be superimposed on old structures for water management. As a mechanism to buy off the old guard in the water sector and induce them to participate in the ongoing process of renegotiating patterns of water allocation and use, the international community took them on study trips abroad and outfitted their offices with new equipment. For example, in the spring of 1994 USAID sponsored a study trip to the United States for 22 water resources managers from the five Central Asian states to familiarize them with the way various US agencies address common water issues.[71] Similarly, in November 1995 the EU-TACIS program organized a study tour to Italy and Germany as part of their project training program (World

Bank 1996b). UNDP also sent overseas representatives primarily from the water ministries and Goskompriroda for training programs on water resources.[72]

More important, the international community rewarded those not sitting at the table; that is, the environmental groups and eco-nationalist movements who were clamoring for the Aral Sea to be saved along with the regional leaders in the disaster zone. Here, the international community took a more short-term perspective and introduced visible projects as a form of side payments. The launching of various development projects to provide for the basic needs of the populations residing in the disaster zone enabled the international community to begin to take over the role of the national governments as the supplier of social protection.

At the inception of independence, the environmental and eco-nationalist groups continued to trouble the Central Asian leaders with their appeals for an end to the system of cotton monoculture and for a concrete solution for the Aral Sea crisis. Yet, without financial and material resources at their disposal, the leaders abruptly turned their backs on the populations in the disaster zone, as vast deserts separated the populations from the capital cities. The Uzbekistani government, in particular, began to attack the eco-nationalist and other opposition movements which had earlier furnished it with a base of support during the glasnost period as to its struggle for greater autonomy from Moscow. The Uzbekistani government took drastic measures to curtail the political and social activities of both Birlik and Erk. The Uzbekistani leaders were more concerned about cultivating cotton in the Aral basin in order to procure much needed foreign revenue; it thus carried out a campaign to squash all remnants of the opposition, most of which ended up in exile.[73]

Facing both a disgruntled opposition and general population, the leaders wittingly agreed to sign a new interstate agreement in exchange for aid to address the Aral Sea crisis. By obtaining a commitment from the international community for assistance, the downstream leaders could then proclaim that they were assuming responsibility for tackling the crisis, while simultaneously curtailing opposition activity. Moreover, the act of engaging in interstate cooperation provided an opportunity to appropriate the language of the eco-nationalist movements who were mounting a real challenge to the authority and legitimacy of the new governments.[74]

One way in which this was done was to create state-sponsored environmental movements that would present the "new voice" of environmentalism on behalf of the government. For example, in Uzbekistan, the state-sponsored ECOSAN (International Ecology and Health Foundation), established in 1992, claimed 5 million members. While I was in Nukus, Karakalpakstan, in August 1994, some members of the Union for the Defense of the Aral Sea and Amu Darya suggested that the creation of this official NGO was to counter the rise of indigenous social movements and for the government to have its own showpiece NGO to present to foreign delegations. The government constituted ECOSAN, in particular, to raise foreign funds for state-controlled programs. For instance, it received a $3.2 million project with UNICEF to provide humanitarian aid to children and mothers of Karakalpakstan.[75] Here, the Uzbekistani government employed its new environmental face as a means to negate the claims of the opposition that it was ignoring the plight of the Aral Sea. Even by appropriating the discourse of the environmental movement, the downstream Central Asian governments still needed to provide tangible evidence that they could meet the needs of their populations residing in the disaster zone in order to maintain an acquiescent domestic constituency. This required sustaining previous patron-client networks in the regions closest to the Aral Sea by making sure that the hokims and akims had sufficient resource flows at their disposal. The Central Asian governments clearly understood that if they supported efforts for interstate cooperation, they would receive assistance at the national level in the form of specific development projects. Many of these hokims and akims were quite influential in shaping development issues in their particular regions. For example, the akim of the Qyzlorda Oblast has organized a regular meeting of development people nearly every 6 months, and most of the donor agencies have attended.[76] UNDP has been his main counterpart in this, and they have encouraged him to set up a water-related subgroup as a result of these activities.

Regarding USAID's early initiatives, it undertook three concrete projects to improve water quality and public health conditions in some of the hardest hit regions near the Aral Sea region—Karakalpakstan, Aralsk, and Dashhowuz.[77] In Turkmenistan, USAID set up a demineralized water treatment (reverse osmosis) plant near Kunya Urgench in Dashhowuz province; in Kazakhstan, it has installed replacement pumps in 32 wells in the area north of the Aral Sea along with the installation of chlorinating equipment at six pumping stations; in Uzbekistan, USAID supplied water-quality improvements for the populations of Urgench in Khorazm province and in Nukus, Karakalpakstan with the installation of chlorinating and chemical equipment at two major water-treatment plants.[78] By providing assistance to all three downstream countries, USAID guaranteed that no one region saw itself benefiting more disproportionately to the others.

The Central Asian governments used this development aid to appease the local hokims or akims and environmental groups who continued to demand recourse to deal with the Aral Sea crisis. The procurement of international aid to the regions hardest hit by the crisis relieved the center of the burden of channeling limited government funds to these areas. For the regional leaders, international assistance helped to solidify their power bases, as they too become dispensers of patronage. Nevertheless, even internally, foreign advisors observed noticeable disputes in the regions between local groups over who will and will not receive certain development projects and assistance. This is especially evident in the places hardest hit by the desiccation of the sea, such as Aralsk or Nukus, where international assistance is still a scarce and coveted resource. The appointed hokims or akims decide how to channel the funds to their constituencies, and thus control a great source of power. Among local groups in Aralsk there was much internal fighting about where the project would take place.[79] By designating the place of a specific project, the hokim/akim could exchange loyalty for the provision of a public good.

UNDP also played a similar role in channeling side payments in the form of technical and financial assistance to the domestic level. Besides having carried out projects at the interstate level to foster a sustainable development convention and to strengthen the capacity of the apex organizations, UNDP embarked on several country-level projects. In Uzbekistan, it focused on human resources through training programs at the national level, the promotion of micro-projects to meet urgent human needs, along with increasing community participation through the improvement of health, education, and social sanitation services. Specifically, UNDP provided direct assistance for a micro-credit program and for the installation of several hundred hand pumps in Karakalpakstan. Momir Vranes from the UNDP emphasized that the small-scale grassroots projects were some of the most popular programs, especially in Karakalpakstan, as the local populations "finally realize that something is happening for their country."[80] Many of these programs were at the initiative of the local population and coordinated with other Western NGOs. Here, UNDP worked with the Mercy Corps on a micro-credit project. Similar projects have been underway in Kazakhstan in Qyzlorda, Kazalinsk, and Aralsk, all located in the Aral disaster zone.

In a similar manner, Israeli development workers introduced water-saving technologies like drip irrigation to increase crop yields while cutting water consumption on pilot farm projects throughout Central Asia.[81] Agridev from Israel in conjunction with USAID carried out several of these projects in which they are trying to acquaint the Central Asians with the concept of the "family farm." In Uzbekistan, the representative from Agridev worked directly with the Ministry of Agriculture in the farm selection process. This is another example of how the government allowed for patronage to flow to those with vested interests in the agricultural sector. On many of these farms, the general population had not been paid for months; at the same time, the heads of the farms were profiting through resources from abroad. On one visit to an Agridev project on a state farm outside Almaty, Kazakhstan, I witnessed the absolute authority of the head of the state farm. While a few local members were building a new dairy as part of a USAID/Agridev assistance program, the head of the farm was selling several dozen cows at his own discretion, without consulting others on the farm, including the head of the dairy.

On the whole, the Central Asian governments allowed multiple international actors to serve as dispensers of compensation in the form of financial and technical assistance. These external sources of resources enabled national elites to placate the short-term interests of their regional clients in exchange for short-term payoffs of political and social stability during this transitional period. Outside compensation permitted the Central Asian leaders to put off sweeping economic reforms that could improve economic and environmental efficiency.

Alongside IOs and bilateral assistance programs, Western NGOs play a prominent role at the domestic level. NGOs fill a critical space that governments and IOs often overlook in the domestic arena. Unlike IOs that seek to influence the high-level negotiations and decisions over water management policies and institutions, NGOs operate in an untraditional manner when measured against conventional understandings of mediation and intervention. In the discussions over new interstate water institutions, many local and Western NGOs, however, expressed dismay that they have rarely been consulted.[82] Because the Central Asian elites made many of the crucial decisions in conjunction solely with IOs such as the World Bank, many local Central Asian groups were excluded from participating in the bargaining process over the nature of both the new interstate institutions and the new domestic institutions.

As a result, Western NGOs aided these societal groups left out of the bargaining process over restructuring the institutions for water use and allocation. They have done so on three levels. First, Western NGOs advocated to have the marginalized voices heard in the discussions on interstate water-sharing patterns and environmental protection in Central Asia. Second, they initiated concrete projects at the local level, focusing on local patterns of water use, health issues, and agricultural practices that directly affect communities living near the Aral Sea. Here, Médecins Sans Frontières has implemented numerous health projects in the region such as dealing with the tuberculosis epidemic.

Third, Western NGOs along with the UNDP reinforced local NGOs as the basis for the emergence of a civil society and to promote an environmental consciousness, both locally and globally. ISAR's Seeds of Democracy Project funded by USAID dispensed small-scale grants to assist environmental groups with institutional development, administrative support, and ecological projects. In contrast to the large-scale, multilateral technical and infrastructure projects, Western NGOs and the UNDP believed that strengthening local environmental NGOs such as the Dashhowuz Ecology Club, the Union for the Defense of the Aral Sea and Amu Darya in Nukus, Green Salvation in Almaty, Ekolog in Toshkent, For an Ecologically Clean Fergana, and Perzent in Nukus was necessary for supporting regional cooperation, civil society, and sustainable economic development.[83] At the same time, NOVIB established an Association of

Aral basin NGOs (ANPOBAM) uniting more than 25 Aral basin NGOs to lobby national and international organizations to reform the principles of water usage in the Aral basin and to increase awareness among the local population on health issues. The first meeting took place on November 22–24, 1996 in Nukus, Karakalpakstan; a second was held on May 16–19, 1997 in Fergana, Uzbekistan.[84]

Another notable NGO is the Aral Sea International Committee (ASIC), which has fought to make sure that many of these local NGOs were represented at various workshops sponsored by the World Bank and at the donors' conferences. One of their main success stories according to its head, Bill Davoren, was getting the only NGO representation at the two main "Participants' Meetings"—the first one was the donors' meeting in Paris in 1994 and the second one took place in Toshkent in October 1997.[85] Not surprisingly, the water nomenklatura compete with the local NGOs for much of the available international assistance, and ICAS often sees itself as the "gatekeeper" to the international community. At the October 1997 donor's meeting, ICAS at first chose not to invite any of the local NGOs, but with much persistence ASIC, the Union for the Defense of the Aral Sea and Amu Darya, and Médecins Sans Frontières attended the meeting.[86] Thus, with the support of Western NGOs like ASIC and ISAR, local NGO leaders have built up their own programs and sustained their activities.

From the above discussion on the active role of IOs and NGOs, it is possible to argue that the breadth and scope of their activity contradicts the claim that Central Asian governments preferred to maintain a system of social control and to limit and/or control the pace of economic and political reform. Why would Uzbekistan and Turkmenistan, the least democratic countries in Central Asia, allow Western NGOs to operate at the domestic level, insofar as they were still interested in cultivating cotton and in maintaining the system of social control based on reciprocal patronage relations? IOs, bilateral aid organizations, and Western NGOs assumed an enlarged role because they began to replace Moscow as a dispenser of patronage to the regional levels. The governments could no longer afford to pay the costs of the social sector in exchange for acquiescence, and as a result, they invited IOs and Western NGOs to intervene, albeit in a highly controlled manner. Since the regional leaders are still appointed, their position of power is contingent on their loyalty to either President Niyazov or President Karimov, for example. By allowing international assistance to flow through the regional leaders, they are able to procure additional revenue outside the limited funds they receive from the budget of the central government. In this area, ISAR installed a wind generator at a maternity hospital in Aralsk, Kazakhstan that was only receiving electricity approximately 50 percent of the time, even though it was on the power grid for the region.[87]

Moreover, by encouraging Western NGOs to operate in the disaster zone, the Central Asian leaders can showcase to the international community that they are embarking on a transition away from authoritarian rule toward democracy. Indeed, the rise of local NGOs provides a good measure for the development of local civil society. The Central Asian leaders also recognize this, and as a result, have sought to co-opt local NGO activities and only allow them to have an environmental and educational component, rather than a political one. While Western and local NGOs are shouldering the costs of providing for social protection, they are still not allowed to agitate and work completely unhindered. For example, NGOs must be officially registered with the government in order to procure foreign funding. Oleg Tsaruk, a member of both the Law and Environment Eurasia Partnership (LEEP) and Aral Sea International Committee pointed out that most Western NGOs work only with officially registered organizations, and the process of becoming officially registered is very difficult and expensive.[88] The Kazakhstani government has placed numerous legal and political constraints on local NGOs by restricting NGO participation in political activity; as a result, the large majority of NGOs in Kazakhstan shun direct confrontation with the government (Jones Luong and Weinthal 1999).[89] Furthermore, most of the local NGOs listed in the UNDP Aral basin Directory focus on environmental, health, or educational activities, rather than on policy making. Overall, these local NGOs do not act as a form of opposition to government policies as did the eco-nationalist movements under glasnost.

Yet, by working with local NGOs, Western NGOs are helping to pick up the pieces after the collapse of the Soviet Union. NGOs in contrast to IOs and bilateral assistance organizations often fill the void of a lack of civil society that remains a vestige of the communist legacy. They do so by focusing on community participation and programs that promote self-reliance.

Conclusion

The combined efforts of IOs, bilateral assistance organizations, and NGOs provided the Central Asian government with much-needed financial and material assistance after the Soviet Union's collapse. The deployment of these side payments helped to foster regional cooperation and to deal with the simultaneous transitions confronting these weakly institutionalized states. Indeed, the cooperation problem facing the Central Asian states over their shared water resources could only be explained within the broader process of state making. Although the Central Asian states achieved rapid regional cooperation, the legacy of cotton monoculture limited the scope and form of cooperation that emerged. More important, it undergirded the shape of the new domestic institutions, especially in the downstream states that were most dependent on cotton revenue.

7

Reconstructing Cooperation in the Aral Sea Basin: Adding and Subtracting Sectors[1]

Despite several setbacks at the end of the first phase of the World Bank and European Union's Aral basin programs, the international community's involvement in the institution-building process did not dissipate. In early 1997 the World Bank concluded its Global Environmental Facility (GEF) appraisal mission, which allowed it to launch the GEF project in September 1998 to stabilize the Central Asian environment and improve the management of the international rivers in the Aral basin (World Bank 1998). After the completion of "WARMAP-Phase 1," in June 1997, the European Union renewed its commitment to support the interstate institutions for water management. Despite having re-tendered the second phase of its WARMAP project, the European Union permitted the legal component of WARMAP to proceed unhampered in order to finish drafting several supplemental interstate water agreements. At the same time, other third-party actors were trying to expedite the process of building interstate institutions in the Aral basin. As of late 1996, USAID redirected its efforts in Central Asia to embark on a limited water-sharing agreement over the Syr Darya River, and in March 1998, the Syr Darya riparians reached an agreement over water releases from the Toktogul Reservoir.

Whereas the previous chapters emphasized how third-party actors facilitated cooperation through side payments, this chapter assesses why cooperation took certain forms and not others in the Aral basin. The process of building interstate institutions manifests cooperation in Central Asia; yet, these institutions vary across scope and form. The lure of side payments pushed the Central Asian states in the direction of creating new interstate agreements for the Aral basin; however, the institutional and accompanying organizational arrangements that ensued were not the only viable solutions to the problem of Central Asian cooperation. In fact, the Central Asian states in conjunction with the international community could have constructed the solution to the Aral Sea crisis along two other dimensions depending on which sectors were included in the institutional agenda. In short, three different sets of negotiations were available to the Central Asian states at independence: a water set; a water and energy set; and a water, energy, and agriculture set.

By early 1998, most international programs had converged along the first two paths contrary to the efforts of NGOs and local activists at the end of glasnost to halt cotton monoculture. Albeit the political and scientific recognition that water withdrawals for cotton cultivation were directly responsible for the desiccation of the Aral Sea, the Central Asian leaders chose not to concentrate their reform efforts on restructuring the agricultural sector. Rather, the Central Asian leaders excluded the agricultural sector in the feasible set of solutions to the Aral Sea crisis for political reasons even when its incorporation at the outset of independence could have produced the most economically and environmentally efficient results in the long-term. The Central Asian leaders were rational actors whose primary preference was the preservation of their power bases. They realized that the costs of including agriculture in the institutional equation for mitigating the Aral Sea crisis were too high in the short-term. To deal with agricultural reform, the Central Asian leaders needed to compensate or displace a large number of vested interests engaged in cotton production. For Turkmenistan and Uzbekistan, the costs were even steeper since they would have to forfeit a tangible percentage of their foreign revenue while lacking another immediate viable export commodity.[2] As a result, the Central Asian elites compartmentalized the water, energy, and agricultural sectors in their negotiations with the international community.

This chapter returns to the theme of cotton monoculture as a system of social and political control in order to explicate why the Central Asian leaders followed a strategy that was the least preferred alternative for ameliorating the desiccation of the Aral Sea. The chapter's main finding is that the desire to maintain political and social control overrode all initiatives to mitigate the Aral Sea crisis.

Constructing the Negotiating Set: Adding and Subtracting Sectors

Issue linkages are an important element in bringing about international environmental agreements since, like side payments, they help to balance the asymmetry of interests and capabilities among the parties to the negotiations. Linking issues, or what is referred to as negotiation arithmetic, creates different options for building institutions for cooperation by increasing the opportunities for tradeoffs and for generating mutual benefits to trade (Susskind 1994; Sebenius 1983). For issue linkage to play a constructive role in any negotiations, it helps if a third party or a mediator can play a role as a neutral actor (Susskind 1994).

In the years immediately after independence, side payments were sufficient to compensate those actors in the water sector who could potentially undermine interstate environmental cooperation. As upstream riparian interests began to crystallize, third-party actors, however, needed to broaden the scope of the negotiations to involve other sectors beyond water and the environment. They needed to focus on appeasing states like Kyrgyzstan that wanted to exploit the upper watershed for hydroelectricity, refuting the downstream users' claims to water rights. At the same time, they needed to take into account that Uzbekistan and Kazakhstan sought to ensure that they would receive the same quantity and quality of water as they historically had for irrigation and agricultural purposes.

Although no violent conflict took place over natural resources since independence, by 1996 it was evident that these new institutions were not the most suitable and efficient ones for addressing the root causes of the Aral basin crisis and for dealing with new sectoral conflicts at the interstate level. Indeed, the introduction of political borders between upstream and downstream users created a real potential for discord as to whether to operate the upstream installations according to the previous irrigation regime or to shift their primary usage to hydroelectricity production. The Soviet water system, that linked the agriculture, water, and energy sectors across the five Central Asian states with the sole purpose of ensuring sufficient water for irrigation downstream during the summer months, was falling apart.

If the international community were to promote a policy of issue linkages along with side payments in Central Asia, it needed to consider the extent and degree to which expanding the agenda or the number of actors would actually strengthen interstate cooperation in the Aral basin. While adding sectors can widen the zone of possible agreement, subtracting sectors can make implementation more manageable. If a third party could encourage the Central Asian participants to add another sector to the negotiating set, this might bring additional parties to the bargaining table. In turn, this could potentially help the Central Asians to overcome the main vested interests impeding the implementation and strengthening of institutions for cooperation in the Aral basin. For example, during the original water negotiations immediately after independence, the Kyrgyz Energy Holding Company was never approached. Yet, at the same time that issue linkages can bring reluctant domestic parties to the table, as the agenda is enlarged, additional domestic interests will want to be consulted (Susskind 1994). Bearing in mind that side payments were critical for inducing cooperation in the Aral basin immediately after independence, third-party actors in subsequent negotiations had to take into account that the widening of the bargaining space could result in expectations on the part of the Central Asians to provide side payments to a larger number of actors than realistically feasible.

Third parties assume an enormous role in deciding who can and cannot join the negotiations. In Central Asia, however, there was not just one third-party actor or transnational actor willing to intervene, but rather multiple parties seeking to intervene and design the institutions for cooperation. Initially, the World Bank and European Union focused largely on water and the role of side payments in the creation of new interstate organizations for resolving the Aral Sea crisis. Yet at the end of the first phase of their programs they wanted to narrow the number of water and environmental issues and parties involved in order to make the Aral Sea Basin Program more manageable. In contrast, USAID broadened the number of actors and issues on the agenda and thus embarked on establishing a joint water and energy negotiating set. However, the major international organizations and bilateral assistance organizations chose not to engage in a three-prong (water, energy, and agriculture) negotiating set despite the wishes of the eco-nationalist movements and NGOs to reduce the amount of land devoted to cotton cultivation as the means to save the Aral Sea.

Negotiating Set 1: Water

Between 1992 and 1997, the World Bank and the European Union undertook a number of substantial steps to influence water management practices in Central Asia. With conflict prevention as a main objective, chapter 6 illustrated how the World Bank devised a water negotiating set in which the international community provided side payments to those with vested interests in the water sector. Yet by the time that many of these donor-driven projects completed the preparatory phases of their respective programs, the optimism that followed their rapid intervention began to wane. During the first stage of phase 1, the World Bank realized that its Aral Sea Basin Program was progressing "slower than was envisaged [at its formal inception in 1994] and than was intended by the participants in the program" (World Bank 1997, p. 1). The donors anticipated that the preparatory phase of the program would only take a year and would conclude with fully prepared projects by the end of 1995, but instead this was not achieved until the end of 1996. In fact, several donor recommendations to ensure that these new institutional arrangements were not just "on paper" were never carried out. For example, the Central Asian leaders failed to appoint a permanent chairperson for the EC-ICAS, to conclude an intergovernmental agreement recognizing the international status of the interstate organizations, and to specify the mandates and jurisdictions of the new apex organizations (World Bank 1996a). The international community (WB, EU-TACIS, and USAID) was becoming increasingly frustrated that ICAS had not clarified its status and that of its accompanying organizations (IFAS, SDC, and the BVOs). IFAS, furthermore, was unable to fulfill its primary objective to act as a fund for the Central Asian states since after several years, it had still failed to collect the money promised to it by the Central Asian states.

As the WARMAP project was winding down, the task manager adduced that the time factor was particularly crucial to establish trust between the international community and their domestic counterparts, and this was largely the reason many goals were not met within the anticipated timetable.[3] In fact, early on the WARMAP project was stunned by the hostile reaction it received to a legal report recommending ways to strengthen the interstate water institutions.[4] In the absence of experience with foreign consultants, a great deal of suspicion existed on the part of the Central Asians, and only with time were the Central Asian participants and the foreign consultants able to develop a productive working relationship. The first phase served as a learning process in which international actors found they could not impose solutions and institutions on local actors while local actors discovered, in turn, that they could not continue past practices unhindered if they hoped to attract financial and technical assistance.

In spite of the inability of the Central Asians to meet many of the international community's recommendations, the World Bank, nevertheless, noted several signs of progress. Most important, the Aral Basin Working Group from Project 1 concluded a version of a Water Resources Management Strategy. Each of the five independent Central Asian states produced a report titled "Basic Provisions for the Development of the Nation Water Management Strategy" that was synthesized into a collective report titled "Fundamental Provisions of Water Management Strategy in the Aral Basin."[5] The culmination of a water strategy met the World Bank's initial demand that to receive international assistance in a transboundary water issue, riparian states must demonstrate a willingness to engage in joint planning. Moreover, the conclusion of a regional water strategy provided the basis for obtaining future GEF funding.

The first phase of international intervention treated the Aral basin problem as purely a water issue, under which the World Bank and the European-TACIS coordinated their projects largely with the water nomenklatura. Although the Aral Sea Basin Program accomplished its two main objectives—to mitigate the outbreak of water conflicts and to develop a common approach to the water sector—it was clear that institutions for cooperation remained limited and under the control of the water nomenklatura. The international community and the Central Asian participants did not deem it necessary to expand the bargaining forum over the Aral Sea to include sectors other than water. Thus, in the proposed Water Management Strategy in the Aral basin, the Central Asian members of the working group wrote:

> It is not [recommended] to locate the water management as part of some of the other sectors of economy, e.g. agriculture or environmental protection. . . . On the contrary the Ministries of Water should be given responsibilities for planning and management of water and nature protection as a single governmental agency

in each country. This will [ensure] that not only water but also demand for water will also be managed. Within the ministries of agriculture and other organizations—there is no interest in reducing demand for water, they are profit maximizers. Minvodkhoz as their partner was always interested in water conservation. (ICAS 1996d, p. 34)

The ICWC and the Scientific Information Center (SIC) were the driving forces behind this new water strategy that excluded other sectors from the Aral Sea Basin Program. Moreover, Rim Ghiniatullin and Victor Dukhovny were at the helm—two major players from the Soviet- and Uzbekistan-based water nomenklatura. The five-country water strategy called for ICAS to become a "water management parliament" that would submit decisions to ICWC for implementation. The ICWC was supposed to "[work] out the main guidelines of [a] common water policy, approv[e] quotas for annual operation of [the] main water sources [with a] special regard for ecological and health requirements, determin[e] annual volumes of water supply to the river deltas and the Sea" and also to "[work] out recommendations for its member states on pricing policy and loss compensation in joint water resources and makes recommendations on the legislative framework of water use" (ICAS 1996d, p. 110). The real authority for decision making still remained under the control of the ICWC because its decisions were supposed to be binding.

At the same time, Dukhovny was also instrumental in creating a new interstate water organization—the Scientific Information Center—that would perform "research and organiz[e] information exchanges among the Central Asian states in technologies and achievements in water management." It possessed responsibility for preparing "reports, recommendations, measures, norms, and rules for discussion and approval by the ICWC and ICAS" (ICAS 1996d, p. 112). In short, the old water nomenklatura managed to capture the new water institutions and reconstruct their roles to fit with the international community's conditions for aid. They moreover precluded the participation of the energy and agricultural sectors in the Aral basin game right after independence. Table 7.1 summarizes the new organizations constituted after independence.

The European Union's legal project likewise coordinated its efforts with SIC and the ICWC. The WARMAP project together with SIC

Table 7.1
Interstate water management organizations in Aral Sea basin.

Acronym	Full name	Date established	Description
ASBP	Aral Sea Basin Program	1994	Central Asian initiative in cooperation with World Bank to address Aral Sea crisis.
BVO	Basin Water Management Organization	1982	Regional organization to monitor and control water allocation and distribution.
EC-ICAS	Executive Committee of Interstate Council for Aral Sea	1993–1997	Originally created to propose projects to ICAS and to implement approved projects.
ICAS	Interstate Council for Aral Sea	1993–1997	Composed of 25 high-level representatives from five states. Developed ASBP projects. Function later assumed by IFAS.
ICKKU/ICKKTU	Interstate Council of Kazakhstan, Kyrgyzstan, Uzbekistan and later Tajikistan	1993	Promote regional economic cooperation.
ICSDSTEC (SDC)	Interstate Commission for Socioeconomic Development and Scientific, Technical, and Ecological Cooperation (also called Sustainable Development Commission)	1995	Focus on promoting sustainable development, composed of fifteen members, comprising three representatives from each state.
ICWC	Interstate Commission for Water Management Coordination	1992	Recommendation-making body composed of respective water ministers.

IFAS	International Fund for Aral Sea	1994	Originally created to finance programs approved by its Board. As of 1997 replaced ICAS in developing and implementing programs within ASBP framework.
EC-IFAS	Executive Committee of International Fund for Aral Sea	1997 (after merger of ICAS and IFAS)	Consists of 10 members (2 persons from each state) plus secretariat.
SIC	Scientific and Information Center of ICWC		Responsible for preparing reports and recommendations for discussion and approval by ICWC and ICAS/IFAS.

completed three draft agreements by April 1997. Under the direction of Dukhovny, the ICWC endorsed these agreements and forwarded them to the national governments with a request that they be considered at the intersectoral level. Only then could formal negotiations commence under the auspices of the European Union-TACIS. These three draft agreements dealt with institutional structure, water use under present conditions, and joint planning.[6]

Not surprisingly, the competing donor-sponsored programs caused confusion among both the donors and the recipient states regarding the overall responsibility for water management in the Aral basin. Taking this into consideration, the World Bank embarked on an overall review of the Aral Sea Basin Program in late 1996 (World Bank 1996a). As a result of the review process, the World Bank recommended that the Central Asian states narrow the scope of the program and find ways to clarify the overlapping functions of the interstate institutions (World Bank 1997). After the first phase of the Aral Sea Basin Program, the World Bank sought to subtract issues from the agenda so that the program's activities could concentrate on "a few, relatively well-defined strategic regional water management issues." In short, the World Bank was hoping that the review process would lead to a "smaller (both in terms of number of activities and financing requirements), more coherent program, managed by a lean but fully capable set of regional institutions" (ibid.). The review mission's recommendations, nevertheless, dealt primarily with changes in the water sector. They proposed that the newly created regional institutions focus primarily on "(i) facilitating the trading of water among states; (ii) the quality of water crossing international borders; (iii) regional (two or more basin states sharing the costs) water development projects; (iv) development of a system of regional system and regional standards for hydrometeorological monitoring and reporting; (v) development of a regional system for control of basin water control infrastructure; (vi) resolution of riparian issues among the basin states; and (vii) data and information exchange on common problems of water resource management" (ibid.). The World Bank's recommendations mirrored the EU-TACIS' conclusions that the Central Asian states eliminate duplication in the roles of the new organizations if they hoped to strengthen their authority and jurisdiction over interstate water management.

Contrary to the international community's objections to the slow pace of institutional change, the Central Asian leaders surprised many of the donors in Almaty, Kazakhstan on February 28, 1997 when they decided to streamline the existing institutions for water management. During this meeting the heads of state resolved many of the above mentioned points of contention and endorsed most of the World Bank's recommendations.[7] They dissolved ICAS and transferred its functions and subordinate agencies to IFAS, which meant that ICWC and the SDC were to be subsumed under the leadership of IFAS. They appointed Rim Ghiniatullin, the former Uzbekistan Water Minister, as Chairman of the EC, while also naming President Karimov of Uzbekistan as its head. They subsequently moved IFAS to Toshkent, Uzbekistan, and turned it into the main apex organization.[8] Lastly, they reduced the amount of fees to be paid from the budget revenues of the Central Asian states. Starting in 1998, Kazakhstan, Turkmenistan, and Uzbekistan were to pay 0.3 percent of their budget revenues and Kyrgyzstan and Tajikistan were to pay 0.1 percent of their budget revenues.[9] (See figure 7.1.)

The consolidation of IFAS and ICAS into one organization delighted the World Bank. The head of the World Bank Mission in Kazakhstan commented, "The World Bank never liked the split between Almaty and

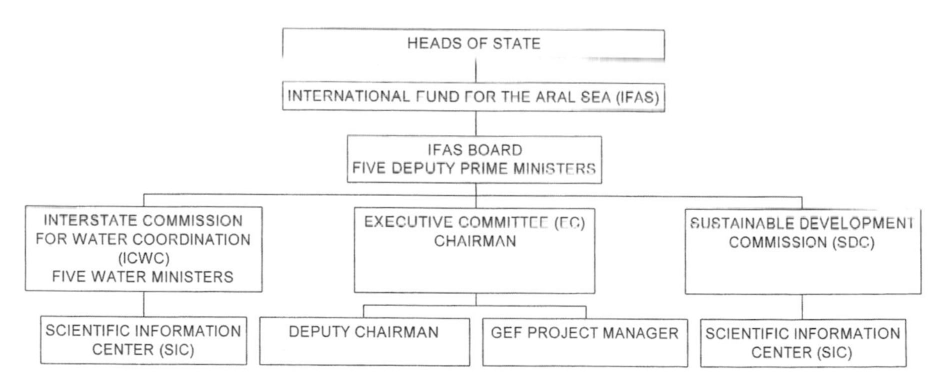

Figure 7.1
A post-1997 organizational chart of the Aral Sea basin (adapted from World Bank 1997).

Toshkent plus a third institution based in Ashgabat, but the World Bank had recognized the politics of the region in which each country got a piece of the program."[10] Many donors tried to speculate why the Central Asian leaders chose to merge the interstate organizations. One donor noted, "Tradeoffs were made in the back room, which had nothing to do with the Aral Sea, but rather to reach an agreement over the Caspian Sea between Kazakhstan and Turkmenistan." Another consultant from the WARMAP team remarked that it is impossible "to dictate a solution, but you can talk about it over and over, and after four times, it comes back as their concept."[11] That seems to capture best what transpired. Thus, the decision to merge the organizations and to endorse the World Bank recommendations appeased the international community that had become impatient with the slow pace of change. More important, the international community had made it clear to the Central Asian participants that they risked forfeiting additional aid.

By constructing the Aral Sea problem as a water negotiating set, both the World Bank and European Union propped up the water sector. Regional cooperation depended on the international community extending side payments in the form of financial and technical assistance to the newly created interstate organizations. However, in the end, these new organizations were, indeed, grafted onto the previous existing networks of the water personnel. According to one international consultant, "Some of the [interstate] structures would have fallen apart owing to their lack of income since these organizations would not have been able to pay salaries. The legal processes and agreements are indirectly driven by the international community [as] people there still think in a command economy structure [and hence] lack international legal experience."[12] In short, the form and scope of cooperation that emerged in Central Asia immediately after independence was a direct result of international intervention and linked to standard World Bank operating procedures in which "the World Bank regulations mandate cooperation," which can take the form of agreements or a water commission.

The resident head of the World Bank's Aral Sea Basin Program in Toshkent described the success of the international community's activities as having "kept people talking. When push came to shove, they never fought."[13] Despite this achievement, the World Bank and the European

Union with their water negotiating set were unable to tackle new kinds of disputes associated with the upstream states redefining their national priorities. The main problem was that the ICWC's authority did not extend over the Toktogul Reservoir since the head gates of Toktogul historically had been under the control of the Kyrgyz Ministry of Energy. If Kyrgyzstan wanted to run it for energy purposes, the BVOs operationally could not prevent this. Hence, they had no real control over the upstream water flow. The water agreements excluded mechanisms for dealing with disputes across sectors, which meant that the Central Asians states had to find other creative means to deal with emerging sectoral conflicts. Moreover, 5 years after independence, this water negotiating set proved insufficient for adequately addressing the environmental and health aspects of the Aral Sea crisis. Many Central Asians were claiming that visible changes to the level of the Aral Sea and improvements in health conditions near the Aral Sea were negligible.

In summary: During the first phase of international intervention in the Aral basin (1992–1997), the form and scope of cooperation reflected the Central Asian water nomenklatura's preferences for dealing only with the water sector. Cooperation was constructed to serve the political goal of ensuring stability and preventing conflict rather than introducing sweeping reform to get at the root causes of the desiccation of the Aral Sea. The international community did not challenge the role of the water nomenklatura in directing policy. Thus, in this instance, security took precedence over environmental protection.

Negotiating Set 2: Water and Energy

The rise of competing sectoral interests seeking to enter the bargaining space made negotiating new institutional arrangements and, for that matter, even maintaining previous institutional arrangements much more complicated. As the first phase of international intervention ended, it became apparent that most members of the international community had overlooked opportunities to broaden environmental cooperation and simultaneously foster a situation of regional interdependence that was characteristic of Central Asian economic and political relations during the Soviet Union. When negotiations focus solely on water sharing, upstream and downstream differences are reinforced, which makes the water gains and losses more prominent (Waterbury 1994, p. 42). The way in which the World Bank framed the problem as purely a water negotiating set ossified stark upstream and downstream disparities among the Central Asian riparian states.

On account of the World Bank and subsequently the European Union propping up the former Uzbekistani water institutions, upstream states such as Kyrgyzstan perceived that their interests were underrepresented in the process of building new interstate institutions for cooperation in the Aral basin. After the February 1997 meeting in Almaty, the World Bank representative in Kazakhstan said: "Once again the Kyrgyz were marginalized by this meeting. Their candidate was voted down [even though] he had been ready to take over since last summer."[14] Thereafter, representatives of Kyrgyzstan's parliament (the Jogorku Kenesh) sent a letter to Michael Rathnam, head of the World Bank's Mission in Kyrgyzstan, asking the World Bank to consider Kyrgyzstan's demand to receive payments from Uzbekistan and Kazakhstan for water deliveries from the rivers originating within their territorial borders.[15] Kyrgyzstani parliamentarians argued that the status quo 1992 water agreement did not reflect Kyrgyzstan's economic interests, but rather only Uzbekistan's water gains. Kyrgyzstan lacked the financial means to maintain the upstream installations that ensured water deliveries to the downstream fields and, thus, appealed for Uzbekistan and Kazakhstan to contribute to the their upkeep.

With respect to the Central Asian leaders' efforts to promote state building (i.e., enhance domestic sovereignty), the Soviet legacy of economic interdependence could have served as an asset for expanding environmental cooperation and regional peace. The Soviet legacy of regional economic specialization fostered a unique relationship of interdependence between water and energy that was overlooked during the first phases of many of the international assistance programs. For example, the division of economic labor in the energy sector during the Soviet Union meant that some republics specialized in supplying energy in certain forms to other republics while other energy sources remained untapped. Kazakhstan possessed a vast supply of unexploited oil reserves but relied on Uzbekistan for natural gas; Kyrgyzstan had the potential to develop hydroelectric power, yet was dependent on Kazakhstan for coal and oil and Uzbekistan for gas; and Uzbekistan had consistently been a large supplier of natural gas to its neighbors but did not control the headwaters of any of the rivers that traversed its territory.

The breakup of the Soviet Union forced the Central Asian states to reexamine these previous patterns of energy use and distribution in order to carry out policies that would forge their newfound independence from one another. Like the Soviet legacy of integrated water management, the Central Asian states confronted the problem of mutual energy dependence while seeking to maximize domestic sovereignty. The Central Asian states confronted a parallel energy game on top of a water game. They needed to devise regional institutions for energy cooperation and to create regional institutions for the management of their shared water resources simultaneously.

Yet, even while the World Bank and European Union were concentrating their efforts on sector by sector analyses, the Central Asian states were still upholding many of these interdependencies by linking water with energy. Kazakhstan, Kyrgyzstan, and Uzbekistan were negotiating annual barter agreements between fuel and water resources. According to these barter arrangements, Kyrgyzstan supplied both Uzbekistan and Kazakhstan with water during the summer months in return for gas and coal, respectively, during the winter months.[16] In addition, Uzbekistan and Kazakhstan agreed to buy from Kyrgyzstan the excess energy generated during the summer months by operating Toktogul. Especially during the water-short years 1995 and 1996, these domestically driven barter exchanges demonstrated the Central Asian governments' determination to minimize interstate tension even though each wanted to pursue a strategy of energy independence.

Since the Central Asian leaders followed sovereignty-enhancing strategies, these barter agreements were unable to prevent minor disagreements from escalating into tit-for-tat disputes over interstate trade in fuel and water resources. Both domestic and international forces were pushing the Central Asian states into conflict with one another. Domestically, the formulation of independent policies in the energy sector by both the upstream and downstream states unintentionally undermined their attempts at water cooperation. With their economies in disarray, each newly independent Central Asian state began to charge world market prices for commodities inherited with an export potential. Turkmenistan and Kazakhstan sought foreign buyers for their gas and oil reserves. Uzbekistan began to sell its cotton harvests on the world market, and in late 1995, it even asked Kyrgyzstan to pay hard currency for gas deliveries. As one of the poorest of the former Soviet republics, Kyrgyzstan could not meet these payments and accumulated huge payment arrears, which prompted Uzbekistan to shut off periodically or reduce the amount of gas delivered to Kyrgyzstan.[17] Without another viable energy source, Kyrgyzstan began to operate Toktogul in the winter months to generate electricity for heating in spite of the barter agreements. Under normal circumstances, the water was stored for downstream usage during the summer months, but instead, the production of hydroelectricity in the winter months depleted the upstream reservoirs, creating a situation of water scarcity downstream in the summer months. Insofar as the ICWC and the BVOs lacked jurisdiction and hence control over the upstream hydroelectric installations, they could not prevent Kyrgyzstan from exercising its absolute sovereignty for independent decision making over water resources originating within its borders. In short, all the Syr Darya riparians including the Aral Sea obtained a sub-optimal outcome even where it appeared that sovereignty-enhancing strategies to achieve energy independence were the most rational.[18]

At the interstate level, the international community also contributed to the rupture in interstate water relations. The international community was supporting privatization projects in Kyrgyzstan and Kazakhstan's water and energy sectors as part of their efforts to promote a market economy. Yet, while the international community was working on the privatization of the electricity sector in Kyrgyzstan and Kazakhstan, it was simultaneously encouraging Kyrgyzstan to abide by the new water-sharing agreements in which it would not generate hydroelectricity in the winter months. Kazakhstan followed a strategy of privatization in its energy sector to raise much needed foreign revenue to compensate for the loss of resource flows from Moscow (Jones Luong and Weinthal 2001). As part of its energy privatization program, it sold its state-owned coal resources in Karaganda. However, in the midst of the privatization sale in 1994, the Kazakhstan government "forgot" to inform the new buyer that an agreement existed between Kazakhstan and Kyrgyzstan in which Kyrgyzstan received coal in exchange for allowing water to flow downstream during the summer. Privatization efforts in Kazakhstan, therefore, disrupted previous patterns of barter exchange between Kyrgyzstan and Kazakhstan in which Kazakhstan bartered coal for water.

If the international community had considered a strategy of tradeoffs across sectors instead of just approaching the Aral Sea crisis as a water problem, this might have reduced the uncertainty of the transition and encouraged continued reciprocity among the Central Asian states. By breaking down the former interdependencies of the Soviet system as part of the state-building process, the international community and the Central Asian governments failed to use to their advantage the Soviet legacy of interdependence in their constructions of the Aral basin problem. As a result, they did not realize the mutual benefits of trading energy for water.

When it became clear that Kazakhstan was unable to deliver coal to Kyrgyzstan, the USAID-EPT project offered its assistance in late summer 1996 to the Interstate Council for Kazakhstan, Kyrgyzstan, and Uzbekistan (ICKKU). The ICKKU, created in 1993 to foster economic cooperation, agreed to US mediation and formed a working group called the Water and Energy Uses Roundtable composed of one water official and one energy official from each Syr Darya riparian state. USAID was hoping to break the impasse among the Syr Darya states, which were having to renegotiate the barter exchanges between water and energy every year.[19] It also wanted to avoid working with the entrenched water nomenklatura, which according to its view comprised the ICAS and ICWC.

USAID aimed to redefine the scope and form of environmental cooperation in the Aral basin by dealing only with the potential and real disputes over water management schemes for the Toktogul Reservoir. Small steps in a limited area such as Toktogul could serve as a confidence building measure to deal later with the broader Aral basin. USAID's decision to render assistance to the ICKKU was premised on the assumption that the 1992 water-sharing agreement was not fixed and complete. USAID justified this shift in strategy because "ICAS was a mess."[20] Michael Boyd, an environmental consultant from the Harvard Institute for International Development in Kazakhstan remarked that the World Bank's program was "doomed to failure by using old mechanisms and institutions created after the collapse."[21] With respect to USAID's decision to overlook ICAS and IFAS, he said this: "Because these are old water management people, environmental issues do not make a dent on these people. . . . Whereas, the World Bank and European Union started with a global solution. They could connect with the old actors who knew how to manage a global system in the old mechanisms by receiving approval or commands from Moscow. Once decisions were made in Moscow, this old guard could manage them. But the old guard cannot manage and do not understand how to deal with new states and markets."[22]

Undoubtedly, the old water nomenklatura sought to restructure the system of water management just enough to ensure that they retained their positions of power and to procure much coveted foreign assistance. However, it is important to recognize that in 1992 the international community (e.g. the World Bank) might not have played such an instrumental role in fostering cooperation and preventing acute conflict if it had excluded the old water nomenklatura associated with ICAS and ICWC at the onset of international intervention. The local water nomenklatura possessed most of the knowledge after the breakup of the Soviet Union regarding the water system and, as a result, had the ability to subvert early international efforts for building regional environmental cooperation. Not having experienced a real independence struggle or revolution that might have wiped out previous institutions, the Central Asian leaders lacked a clean slate upon which to create new water institutions. Since the Central Asian republics only became states through the process of state breakup, the former nomenklatura continued to assume an influential role in directing the transition away from state socialism.

In order then to develop a multi-year agreement for the operation of the Toktogul Reservoir, USAID sponsored the first meeting of the roundtable (October 1 and 2, 1996). At this meeting, the participants discussed the timing of releases and optimization of water uses from Toktogul, energy prices, cost sharing for infrastructure and maintenance, damage estimates, and the impact of privatization on the ability to meet agreement provisions.[23] A second meeting of the Roundtable was held on December 18–20, 1996 in Almaty, Kazakhstan. This time US experts presented different models for running Toktogul to demonstrate the mutual benefits that could be obtained from maintaining an irrigation regime if compensation was included.[24] In addition, USAID circulated a draft treaty for the

Syr Darya to serve as a framework for a multi-year interstate agreement. Discussion continued at the third session (July 1–4, 1997, Ysyk-Kol, Kyrgyzstan). Here, the participants concentrated on developing the notion of compensation arrangements for wintertime water storage and summer releases from the Toktogul Reservoir.[25] The Central Asian representatives agreed in principle to the form and extent of interstate exchanges of fuel and water resources.

At the fourth Roundtable (Almaty, September 7–12, 1997), USAID continued its work on producing a multi-year interstate agreement. USAID hoped cooperation would be based on yearly negotiations over water and energy use on the Naryn-Syr Darya cascade of dams, which would enable the Central Asian states to regulate yearly flows into the Aral Sea. USAID again circulated a draft agreement that would cover 5 years. It included provisions for energy and water exchanges, government guarantees for these energy exchanges, either in energy or currency reserves, and payment of operation and maintenance costs. It also proposed a new umbrella organization to "bypass the ongoing squabbling between existing regional organizations seeking control and management of the Syr Darya and the associated water rights."[26] At this meeting, the ICKKU member states recognized that Tajikistan needed to be included in negotiations as the Kairakum Reservoir is located within its territory.

USAID drove the agenda during these roundtable meetings. For example, it presented its version of several draft agreements hoping to persuade the Central Asian participants to establish specific numbers for how much water to deliver downstream in exchange for how much fuel to be delivered upstream.[27] Its efforts resulted in the interstate water compact among Kazakhstan, Kyrgyzstan, Tajikistan, and Uzbekistan signed by the prime ministers in 1998, creating an alternative framework for dealing with conflicts on the Syr Darya.[28] Instead of just the water nomenklatura managing the distribution and allocation of water resources, this agreement called for cooperation between the energy dispatch center (UDC Energia) and the regional water supply and distribution organization for the Syr Darya (BVO). Both their representatives had participated in the above round tables.

Unlike the World Bank and European Union's initial programs, USAID constructed an alternative negotiating set that added the energy sector to the bargaining agenda while simultaneously subtracting the Amu Darya basin from the negotiations. Rather than adhere to an integrated basin approach, USAID broke down the Aral basin into two separate river basins. This gave it more control over the actors involved, and it also removed the most difficult riparian, Turkmenistan, from the bargaining forum. Turkmenistan, of all the riparians, was the most resistant to outside intervention in the Aral basin. Overall, it has followed an isolationist strategy relative to the other former Soviet republics.

USAID concluded that broadening the scope and form of cooperation in the Aral basin could not just depend on side payments, but needed to include issue linkages since Kyrgyzstan's use of hydroelectricity in the winter months was beginning to disrupt the water regime. The addition of sectors allowed USAID to expand the zone of agreement, and the use of compensation of fuel for water created a potential for tradeoffs allowing them to overcome interstate disputes. Yet, by establishing an alternative game to ICAS, USAID had to weigh in more heavily on the side of Kyrgyzstan to equalize the asymmetries between the upstream and downstream riparians. This required paying close attention to Kyrgyzstan's need for energy and supporting Kyrgyzstan's interests when setting the interstate policy agenda. USAID could not just look at its Aral Sea Basin Program as isolated from its energy program.[29] In fact, the head of the ICKKU, Bazarbai Mambetov, perceived that the programs were inseparable. In September 1997, Mambetov called for a meeting with the head of the Hagler Bailly project in Bishkek Kyrgyzstan, to persuade USAID to sponsor projects under the ICKKU's auspices that would benefit Kyrgyzstan. He proposed a new program that would "convert the entire Kyrgyz population to electricity for ALL purposes." In short, Mambetov hoped that USAID would support Kyrgyzstan's drive to expand hydroelectricity consumption and hence provide the funding to complete the Kambarata 1 and 2 hydroelectric installations on the upper Syr Darya.[30]

USAID's construction of the Aral basin problem differed from the approaches of the World Bank and the European Union. USAID created a new negotiating forum that was in competition with that under the ICAS's jurisdiction. Moreover, USAID drove the bargaining process by proposing draft treaties and by "paying" the participants in the roundtable meetings.

USAID fostered a situation of patronage, reminiscent of the Soviet system in which the clients were supposed to acquiesce to the directives of the patrons. According to the USAID EPT Regional Water Policy Advisor, "if Mambetov did not want to discuss the various scenarios [proposed by USAID for Toktogul], he would not get paid."[31] Despite having facilitated a new agreement, USAID's decision to work outside the World Bank structure was, nevertheless, controversial. One consultant from the European Union described USAID's policy as "divide and rule."[32]

This alternative construction of the Aral basin crisis as a water and energy problem indicates that if international actors had recognized the linkages between the energy and water sectors immediately, the Central Asian states might have been able to thwart festering conflicts of interests and capabilities over Toktogul. Although coupling water and energy together was necessary to prevent conflicts over water use, it nevertheless would not have been sufficient to improve the environmental quality in the Aral Sea and the near-Aral region. Only a strategy that connected the Aral Sea crisis to agricultural practices would have been both necessary and sufficient to improve the environmental situation within the Aral basin. However, the Central Asian leaders along with the international donor community were unwilling to undertake sweeping reform in the agricultural sector.

Negotiating Set 3: Water, Energy, and Agriculture

Why did the downstream leaders, who had pushed for international intervention to "save the Aral," end up resisting immediate reform in the agricultural sector, especially when it was clear that cotton production was the largest consumer of water in the region? At the same time, why did local activists curtail their demands for reducing the amount of state quotas for cotton? If the real issue at stake in the Aral basin was agriculture, why was it not placed high on the policy agenda when devising interstate solutions to ameliorate the desiccation of the Aral Sea? Indeed, many in the international community recognized the importance of dealing with agriculture. "Water is a symptom of the larger problem between Uzbekistan and Kazakhstan; this concerns what is and should be the size of the agricultural sector in these countries—cotton and food grain," said the resident head of the World Bank in Kazakhstan. The cause of the water problem was that all the Central Asian states "want to be independent in food production as well as to continue cotton production for export."[33]

The push for food security and the maintenance of cotton production precluded sweeping reform in the agricultural sector, and without real agricultural reform, attempts to mitigate the Aral Sea crisis were limited. This section explicates why certain solutions to the Aral basin problem were excluded from the policy agenda after independence. Both the water set and the water and energy set were not the only ways in which the international community could have constructed institutional solutions for cooperation. Rather, a third option existed that would have included the agriculture sector. Under this scenario, the Central Asian states and the international community would have had to replace cotton with less water-intensive crops. In the long-term, this was the most efficient strategy to rectify 70 years of disregard for the environment.

Whereas this option appears the most economically and environmentally rational for mitigating the desiccation of the Aral Sea, the Central Asian leaders perceived that the shift away from cotton monoculture would have politically and socially destabilizing consequences. In the short-term, the linking of water and energy with agriculture might have overconstructed the water problem in Central Asia for reasons to do solely with domestic politics. On the one hand, Turkmenistan and Uzbekistan sought to keep the general population on the farms and engaged in cotton production to ensure their hold on social control and stability. On the other, the leaders of Turkmenistan and Uzbekistan could not jeopardize the foreign revenue earned by cotton sales abroad. Any interstate solution to the Aral Sea crisis would have been futile without the cooperation of Uzbekistan and Turkmenistan since they were the two largest consumers of water in the region, and moreover, their populations were located in the near-Aral region. Simply put, domestic-level conditions illuminate why the Central Asian leaders resisted agricultural reform even though it was obvious that the exclusion of agricultural reform from the Aral Sea Basin Program could undermine institutional efforts to ameliorate the Aral Sea crisis and deepen interstate cooperation.

The first reason why institutional solutions to the Aral Sea crisis omitted agriculture was the Central Asian water specialists preferred "simple" technical solutions to "complicated" political solutions. The notion of a "technical fix" still permeated the water nomenklatura's Weltanschauung when dealing with the root cause of the Aral Sea crisis. Even though the Central Asian participants in the Aral Sea Basin Program recognized that the likelihood of carrying out large water diversion projects was wishful thinking, the Joint Water Strategy produced by Working Group 1 heavily favored technical solutions. Proposed solutions focused on water conservation, which included relining canals, reconstructing the drainage system, installing water meters, and land leveling (ICAS 1996d, p. 91). As during the Soviet period, water planners preferred building and reconstruction projects because these projects brought with them large sources of financial resources. The main project under consideration for the downstream states was the Amu Darya Right-Bank Collector Drain, which was being prepared for appraisal in 2001. The upstream states had also proposed big dam projects on the upper reaches of the watershed.

Second, the collapse of the Soviet Union created a situation in which fragmented economic sectors competed with each other for access to foreign assistance to stay afloat. Owing to the shortfall of domestic rents after independence, the economic ministries scrambled to find potential donor partners. Like the donors who competed with each other for projects in Central Asia, the water and agricultural ministries preferred their own projects. This way foreign assistance would stay under the roof of one ministry rather than having to be spread around several ministries. Accordingly, the water nomenklatura only considered agricultural projects if they were related to reconstructing the irrigation networks to improve efficiency. Similarly, the Ministry of Agriculture in Uzbekistan solicited projects that would raise cotton yields rather than encourage the introduction of small-scale farming through extensive privatization. In response, the World Bank initiated a cotton sub-sector improvement project in Uzbekistan to boost the quality and the supply of seed, elevate cotton grading to international standards, and advance methods for pest control. The Eisenberg Group from Israel introduced drip-irrigation technology on several pilot farms to improve water efficiency in cotton production. The Meredith Jones Company from England later established pilot farm projects to raise the quality of cotton production.

Indeed, each sector sought to protect its own interests immediately after independence.[34]

Third, the economic importance of cotton in Uzbekistan and Turkmenistan continued to climb after independence. Uzbekistan and Turkmenistan decreased the use of land for growing cotton only when that was absolutely required by the salinization of the soil or to make room for food production in some of the most densely populated areas (Craumer 1995, p. 16). In Turkmenistan, for example, cotton remained a "major cash crop" of which half of all irrigated land were dedicated to cotton production alone (World Bank 1994b, p. 130).

Cotton generated a disproportionate share of foreign exchange for both Uzbekistan and Turkmenistan immediately after independence. In 1991, cotton already comprised approximately 84 percent of Uzbekistan's foreign exports, and by 1992 it provided more than three-fourths of Uzbekistan's total export revenue alone (World Bank 1993b, p. 24; IMF 1992, p. 2). The value of Uzbekistan's total exports in 1992 totaled $869 million of which cotton provided $673 million (World Bank 1993b, p. 24). Between 1993 and 1997, Uzbekistan was the world's fifth largest producer of cotton and second largest exporter of cotton.[35] Although gas exports still dominated Turkmenistan's central budget, the amount that cotton exports provided to the state income continued to climb after 1991 (IMF 1994, p. 125). The Central Asian leaders continued to rely on the system of cotton monoculture because the foreign sale of cotton offered a stopgap measure to ease the costs of the transition and sustain their patronage systems. The UNDP concluded that unlike other CIS countries, Uzbekistan was much better able to defer a large decline in GDP largely owing to its large amount of agricultural output. In short, agriculture was the "shock absorber."[36]

Fourth, Turkmenistan and Uzbekistan justified their policies to continue cotton monoculture on grounds of the "strategic" role these crops play in the economy. The Soviet-era system of goszakaz, in which cotton and grain production targets were still set by the state, allowed them to retain cotton as their main source of foreign revenue. The persistence of controls on the marketing and distribution of cotton reflected "widespread concerns in the government [Uzbekistan] about ceding ownership and direction of the agricultural industry to private enterprise."[37] State commodity boards controlled the production and requisition of cotton under which they set artificially low prices at which the government purchases cotton and then resells it for world market prices to foreign countries. Although farmers in Uzbekistan were only required to sell 50 percent of their cotton to the state, in practice they were forced to sell almost all of it to the state because they lacked direct links to the foreign buyers.[38] Likewise, state orders remained in Turkmenistan after independence, and the Ministry of Agriculture possessed the sole authority to market cotton to the world market (IMF 1994, p. 130). Owing to the high value placed upon bringing in the cotton harvest, the Minister of Agriculture's position depended on his ability to meet the set cotton targets. When performance would decline, the government usually changed and/or replaced the Minister of Agriculture and/or the deputy prime minister in charge of agriculture. Moreover, in Uzbekistan, the continuing preference for cotton was not just a decision of the Ministry of Agriculture, but came from the highest levels of the government bureaucracy including the Ministry of Finance, the Ministry of Macroeconomics, and the President's Office.

Finally, preserving full state control over cotton production and sales has not only enabled state leaders in Uzbekistan and Turkmenistan to use the proceeds from cotton exports at their discretion, it also has allowed them to reinforce regionally based patronage networks. To even begin to dismantle the agricultural sector would threaten the very foundation of social control, which was grafted onto reciprocal relationships among the central leaders, the regional leaders, and the heads of the farms. Despite the environmental legacy of cotton monoculture under the Soviet Union, the Central Asian leaders delayed agricultural reform for as long as possible to hold onto a command economy.[39] As a result, no real change took place in Turkmenistan and Uzbekistan's agricultural sectors as a whole, including their state and collective farm structure. In fact, in the productive rural areas where the population density was high, employment in agriculture kept people on the farms and under the control and surveillance of the government.

Turkmenistan inherited "a population largely dependent on cotton production for employment" (IMF 1994, p. 123). According to the Food and Agricultural Policy Advisory Unit of the TACIS Programme of the

European Commission, "Uzbekistan ranks as having one of the highest farm population densities of all post-Soviet states. The Fergana Valley, widely acknowledged as the most fertile of the country's agricultural zones, has an average population per sovkhoz/kolkhoz of 8131 persons, and an average workforce (including pensioners) of 4129 persons. The average size of these farm units is 1515 hectares."[40] If Uzbekistan pursued land privatization under which land were distributed in equal shares as in Russia and Ukraine, 4129 persons on the average farm would share about 1350 hectares, giving an individual land share of 0.33 hectare.[41] Such reform would unsettle an agricultural system in which most inputs, machinery, and irrigation networks were geared toward large-scale farming.

The Uzbekistan and Turkmenistan governments resisted land reform because it would threaten the very social and political stability that the state leaders were eager to maintain. The dismemberment of state-directed agriculture would reduce the amount of foreign currency resources the state received from cotton production. The privatization of land would also lead to a situation in which people would only devote themselves to smaller private plots and create a situation in which the state would be forced to loosen its control of the population.[42] In short, this would break down the system of patronage ties with the local and regional authorities since the population would no longer be required to turn over a certain percentage of their harvest to the state.

Yet, even when Uzbekistan and Turkmenistan made gestures in the direction of land reform, may of the policies only resulted in the replacement of former Soviet structures with new Central Asian names. Real production relations have remained largely unchanged (Craumer 1995, p. 7). The Land Code of Uzbekistan called for the replacement of the kolkhoz and sovkhoz by co-operative structures (shirkhats), which are worked by their members under "family contracts." However, the contracts resemble the traditional "brigade" system introduced under the Soviet Union. These brigades forced "students, children, industrial workers, soldiers, clerks, and drivers . . . to work in the fields picking cotton for a symbolic wage."[43] Most of the shirkhats were, in fact, the same kolkhozes but on a smaller level and were still responsible for meeting government production targets especially for cotton and grain (Ilkhamov

1998). Uzbekistan and Turkmenistan, furthermore, preferred hand-picked cotton, since it was superior to machine-harvested cotton.

Uzbekistan and Turkmenistan, in particular, lacked the appropriate incentives to relinquish their hold on cotton monoculture and to construct a negotiating set that included agriculture. Even though their populations bordering the Aral Sea were bearing the environmental costs of cotton monoculture, any sweeping reform in the short-term would have resulted in substantial losses in their foreign currency holdings. To ensure that enough cotton was picked, these governments relied on a system of social control in which reciprocal relations linked the farm heads to the political elite in the center. Moreover, the flow of side payments from the international community to important domestic interests in the Aral Sea disaster zone bought the leaders social and political stability in the short-term, enabling them to perpetuate this system of cotton monoculture.

Conclusion

As new interests and capabilities solidified, the international community and the Central Asian states were able to reevaluate the scope and form of the existing institutions created to foster interstate cooperation. They did so by adding the energy sector to the bargaining agenda in order to mitigate potential upstream-downstream disputes. However, in order to rectify the environmental crisis in the Aral basin, the Central Asian leaders needed to address the long-term consequences of maintaining cotton monoculture. Yet the short-term interests to maintain cotton monoculture precluded placing negotiating set three on the bargaining table even though this would have had a greater chance in mitigating the environmental crisis in the Aral basin. In the immediate term, the Central Asian leaders were unable to pay the domestic costs of environmental protection since this would have led to a steep decline in foreign revenue from cotton sales. Thus, the Central Asian leaders continued to sacrifice the "revival" of the Aral Sea for the sake of cotton production. Surprisingly, the international community backed these strategies because they also preferred social and political stability to pursing an alternative negotiation set that could undermine the leaders' hold on power.

8

Making States through Cooperation

A Bucket of Water

There is a common expression in Central Asia: "If every foreign visitor coming to the Aral Sea region brought a bucket of water, the Aral Sea would be saved." This saying in many ways captures the duality of the Central Asian environmental experience. On the one hand, the whole process of regional environmental cooperation has been deeply embedded within the international system in which a multitude of third-party actors worked to bring about alternative institutional solutions to the Central Asian water crisis. In fact, rather than demanding that the Central Asian leaders assume responsibility for the destruction of the Aral Sea ecosystem, the Central Asian scientific and general population naively believed that the international community would provide an immediate solution to the environmental and health problems generated by the sea's desiccation. On the other hand, this dependence on the foreign donor community has led to many missed opportunities and bitter disappointments because the Aral Sea crisis has yet to be fully resolved. Especially for the populations living in the near Aral region, the destruction of the Aral Sea ecosystem has served as a constant reminder of the Soviet Union's abuse of the natural environment. The Karakalpaks, for example, have experienced first hand the economic and health costs associated with the loss of a viable fishery and the rise of dust and salt storms. The most worrying indicator is that mothers have refused to breast-feed their newborn babies because their milk is considered to be so overly toxic from the contaminated water they drink. Owing to the aggravating environmental and health situation, the Central Asians living in the near-Aral region frequently complain that international aid and intervention has failed to translate into real environmental, health, and economic improvements. They find instead that they are only left with empty pledges of support for their plight. Thus, nearly a decade after the Soviet Union's political and physical borders began to crumble, many Central Asians have begun to view the role of the foreigner donor community with much skepticism.

Even if each foreigner had brought with him or her a bucket of water or more significantly if the Central Asian states had succeeded in achieving their objective of an inter-basin water transfer, these outside solutions would not have adequately dealt with the underlying causes of the Aral Sea crisis. The real test for the Central Asian states and the international community has been how best to tackle the Soviet legacy of cotton monoculture. The Aral Sea crisis is not just an environmental and health narrative, but rather it is a narrative of Soviet power and social control. Through the lens of an environmental disaster, this book has exposed the political, economic, and social remnants of the Soviet legacy of cotton monoculture in order to understand their grip on Central Asian state building and regional water cooperation. Even when state breakup disrupted previous patterns of traditional rule based on patronage, the legacy of cotton monoculture enabled national and regional elites to maintain a strong hold on state power and social control in the Aral basin. By providing for a system of social control, cotton monoculture managed to impede any radical measures to effectively address the Aral Sea crisis as most water diverted from the Amu and Syr Darya rivers continued to end up wasted on overly salinized and water-logged fields.

Tied to the duality of the Central Asian environmental experience, the Aral Sea water-sharing predicament illuminates both a positive and negative tale of the ability of third-party actors or what are also considered transnational actors to bring about regional environmental cooperation among transitional states. Although the case of the Aral basin supports the trend that IOs, bilateral aid organizations, and NGOs are much more visible and powerful actors on the international stage, their role is still ambiguous. On the positive side, this case shows that cooperation can emerge even under the most inauspicious conditions with inducements from third-party actors. Whereas the conventional theoretical literature on interstate cooperation and the empirical evidence from other international water basins suggested that these newly independent states should have been concerned first and foremost with sovereignty-enhancing behavior and consolidating their independence, the Central Asian successor states actually coordinated their water policies in the Aral basin immediately after independence. By developing an approach to two-level institution building that integrated the two main challenges facing the Central Asian states—state formation and interstate cooperation—I showed that side payments could help transitional states engage in cooperative behavior as a means to enhance state sovereignty and to integrate into the international system of nation-states.

The Central Asian case thus upheld the supposition that cooperation is nested within the state-building process, especially when international third-party and/or transnational actors are involved. This case demonstrated that without an overtly active role for IOs, bilateral aid organizations, and NGOs, the Central Asian states might not have immediately designed new institutions for regional environmental cooperation; rather, other outcomes of discord or non-institutionalization would have transpired. Cooperation through institution building took place precisely because IOs, bilateral aid organizations, and NGOs furnished side payments in the form of financial and material resources to pivotal domestic constituencies during the transition period. In short, they were able to relieve the Central Asian governments of the loss of Soviet resource flows with alternative sources of assistance. The addition of an active role for IOs, bilateral aid organizations, and NGOs not only reflected the increasing interplay between global and domestic politics, but also influenced the way in which cooperation was reached and the form it took.

The lure of foreign assistance resulted in a plethora of interstate agreements directly and indirectly connected to the Aral basin. These agreements have mitigated violent ethnic conflict over scarce resources in the post-Soviet period. Widespread small-scale conflict like that which occurred at the end of glasnost has disappeared. Yet at the same time, the Central Asian states have not achieved the overarching objective to obviate environmental degradation. This may be the case because it is either too early to adequately judge the effectiveness of these agreements or because these new institutions have remained dormant.[1] More likely, the early emphasis on conflict prevention along with political and social stability impeded attempts to foster real structural reform in Central Asia, which has directly affected the likelihood for improvements in the natural environment. Simply put, the emergence of rapid regional institutionalization in Central Asia indicates that state building and security concerns superseded environmental matters in the aftermath of state breakup. This simple observation has not so simple implications for the type of states emerging in Central Asia and their relationships to their domestic populations and the international community of nation-states.

Thus, on the negative side, this hopeful tale of regional environmental cooperation turned into a missed opportunity for the Central Asian states and the international community because of the strength of the Soviet legacy of cotton monoculture. Taking this as its starting point, this chapter concludes with a discussion of the theoretical and empirical implications for state building and environmental protection generated by the process of rapid regional institutionalization over the Aral basin. These findings have broader significance for understanding the enlarged role of third-party actors and the nature of state sovereignty in the post-Cold War context. First, the case of the Aral basin clearly demonstrates that IOs, bilateral aid organizations, and NGOs are bolstering state sovereignty rather than undermining it. Second, at the same time they are reinforcing state sovereignty, these actors are helping new states cultivate a myth of state and nation making. Third, the process of regional cooperation and state formation embedded within IOs, NGOs, and other foreign assistance organizations has produced decoupled institutions in the water sector. Fourth, IOs, bilateral aid organizations, and NGOs should be wary of too much intervention during the formative years of state building since this can hinder the development of reciprocal linkages between state and society that provide the basis for democratic governance. Even small amounts of foreign aid, such as that which was channeled to the Central Asia states, can mold the nature of state-society relations in the formative years of state making. Finally, the Central Asian case holds lessons for other political entities engaged in both the processes of state formation and regional environmental cooperation.

IOs, Bilateral Aid Organizations, and NGOs: Reinforcing State Sovereignty

The empirical evidence from the Aral basin case demonstrates that IOs, bilateral aid organizations, and NGOs have assumed an enhanced role in world politics after the end of the Cold War. Their impact on resolving global issues and influence on the internal functions of new states is greater than initially perceived by state dominated theories of world politics. Similarly, accepted conceptions of sovereignty as the exclusive domain of governments are no longer applicable at the beginning of the twenty-first century since non-state actors are deeply involved in the internal policy making of independent states. In Central Asia international interventions have covered diverse policy issues ranging from military, finance, electoral, and environmental institutions. In all these issue areas, Western IOs, NGOs, and bilateral development organizations have transported Western models of the rational-legal and democratic state to Central Asia. In doing so these IOs, bilateral organizations, and NGOs have eagerly embraced the discourse of "capacity building" as the appropriate means to build and strengthen new states. State capacity is not a new term for scholars of "the state" (Skocpol 1985), but it has become a buzzword in development organizations and international financial institutions.[2] The argument is that without capacity, states are unable to devise and implement effective domestic policy, to maintain control over a bounded territorial space, and to engage in foreign relations with other states. In short, states that lack domestic governance or capacities to penetrate society are considered to be "weak" or "quasi" states. Through active intervention, IOs, bilateral aid organizations and NGOs attempted to preclude the Central Asian states from turning into full-fledged weak states or even worse from deteriorating into failed states like in Afghanistan.

In order to build state capacity and reinforce state sovereignty, IOs, bilateral aid organizations, and NGOs have pushed for new states both to integrate into the international community of nation-states and to create regional blocs. The Central Asian states have not resisted engagement with the international community or signing interstate agreements with each other.[3] They recognized early on that, owing to the loss of resource flows from Moscow, they would not be able to resolve the Aral Sea crisis through independent action. Cooperation with the international community, moreover, enabled them to actually disengage from the Russian sphere of influence and to address other foreign policy concerns that extend beyond water. The act of signing agreements entailed the mutual recognition of each other as viable nation-states. Thus, they adhered to international efforts to generate interstate institutional agreements as a means to fortify political borders of separation and empirical sovereignty.

In addition to water cooperation, the international community bestowed on the Central Asians pseudo-military arrangements as another means to strengthen their domestic sovereignty. Uzbekistan, Kazakhstan, and Kyrgyzstan have wittingly participated in US-led military training programs. With US support, they created the Central Asian Battalion to combat newfound threats from the Taliban movement in Afghanistan and to gain the capacity to protect their external borders. Western IOs and governments have pushed for these types of military cooperation as a way to diffuse Western norms of appropriate military behavior in the international system of nation-states; simply put, they are hoping to inculcate Western military doctrine and to professionalize the Central Asian militaries.

In contrast to the argument that weak domestic institutions are usually inhospitable to cooperation, in the Central Asian water case, the opposite appears to be true. Here, the nature of weak domestic institutions was pivotal for understanding rapid regional cooperation in Central Asia. International actors assumed an enlarged role in Central Asia because these weak states lacked the necessary domestic capacity and resources to mitigate environmental degradation. In essence, they effectively lacked positive sovereignty, and as a result turned to the international community to compensate for this void.

Yet, while the international community sought to build entirely new interstate and domestic institutions, it was often the case that the Central Asians were encasing previous forms of governance into new institutional structures. With assistance flowing in from the international community, the Central Asian leaders did not have to reign in the old water elite responsible for the water management crisis or to dismantle monocrop agriculture. Instead, the Central Asian leaders consented to the superimposition of new international institutions like ICAS and IFAS upon the old Soviet water structure. International acceptance of these new water institutions allowed the Central Asian leaders to build new states within the wreckage of the Soviet system. In summary: The active role of international actors fortified states that were without any historical antecedent as states before December 1991.

Third-Party Intervention and the Myth of State and Nation Making

Integrally related to the assumption that regional cooperation bolsters state sovereignty is that empirically, the Aral basin case illuminates how the process of building interstate institutions for regional cooperation is really about creating a myth of statehood and nationhood. Leaders in new states perceive that in order to become legitimate members of the international system of nation-states, they must ascribe to certain modes of behavior even if in practice, they are merely symbolic. Post-communist state building entailed engaging in acts of compliance with international norms concerning the role of the state in the international system and the form of the internal components of the state at the domestic level. Interstate cooperation was thus a symbolic step toward sovereignty enhancement. Yet these acts often have little substance attached to them.

The reputed "triumph" of capitalism and democracy after the end of the Cold War increased the coercive capabilities of IOs, bilateral aid organizations, and NGOs in the processes of interstate and domestic institution building (Barnett and Finnemore 1999). The collapse of ideology and the delegitimization of an alternative to the socialist mode of development moreover enabled these transnational actors to assume an all-encompassing role in the course of post-Cold War state formation. IOs and Western advisors are more than ever dictating to post-communist or transitional states what steps they should follow to join the international system of nation-states, irrespective of whether the paths followed are compatible with the historical antecedents. Post-communist states are expected to adhere to democratic principles and integrate into the global liberal economic order. In the aftermath of independence, the foreign aid community inundated the Central Asian states and other post-communist states with numerous advisors and consultants with the stated goal to help them build free markets and Western-style democratic institutions. In short, they instruct countries on what specific agencies and domestic institutions to constitute in exchange for aid.

In order to join Western multilateral organizations, the international community, moreover, expects transitional states to improve their human rights record, to undertake "free and fair" elections, and to adhere to global environmental regulations. Membership in the Organization for Security and Cooperation in Europe is linked to a respect for human rights, and as a result, the OSCE exerts pressure on many of the Soviet successor states to safeguard the rights of minority populations within their territorial borders. USAID, specifically, has been involved in a wide range of development assistance programs besides the environment. It sponsored the National Democratic Institute to assist with carrying out competitive election campaigns, Winrock International (Farmer to Farmer) to foster the creation of small-scale farms, and Hagler Bailly to restructure the domestic energy sector throughout Central Asia.

Since the Central Asian states were extremely poor from the outset they had fewer means to resist outside demands and intervention. This is not to imply that the Central Asian leaders did not consider alternative pathways to follow. Besides the Western model, the Central Asian leaders raised the possibility of a Turkish or even Korean approach to economic development and political control. Yet in the end, only the IMF and World Bank could promise sufficient amounts of assistance to enable the Central Asian states to establish concrete borders of separation from Moscow. Kyrgyzstan's dependence on external aid and the lack of exports for foreign revenue opened up the door immediately for international advisors to play a role in its domestic politics. External aid, ironically, thus became a source of international recognition that Kyrgyzstan was no longer a satellite republic of Russia but a viable nation-state that could take part in negotiations with members of the international community.

Although the argument presented here is that transnational actors are helping to strengthen state sovereignty, it should not be overlooked that in order to reinforce state sovereignty, they are indeed violating state sovereignty. Even when IOs, for example, encourage countries to create agencies and sign legal documents by reason of "this is what states are supposed to do," they also infringe on the domestic sovereignty of states by limiting the ability of states to engage in independent policy making. Indeed, the illusion that domestic governments are actually responsible for the creation of new domestic institutions is essential for state building even if these acts could not be accomplished without foreign financial and technical assistance.

Concerning water issues, international advisors try to impose new norms for water management throughout the world, which includes setting up water and environmental agencies similar to Western ones. In exchange for financial and technical assistance, the World Bank, UNDP, UNEP, and the Global Environmental Facility (GEF) expect countries that share a body of fresh water to create a water basin organization and sign agreements for environmental protection. The international community has initiated similar cooperative programs over the Mediterranean Sea, the Danube basin, the Black Sea, and the Caspian Sea. The United Nations has also been actively involved in the creation of a water basin commission for the Mekong which was originally established in 1957 and then re-formed in April 1995 with the "Agreement on the Cooperation for Sustainable Development of the Mekong River Basin. The Mekong basin with its history of active international participation has provided one model for Central Asian water and environmental experts.[4]

After a water commission is constituted, international advisors and consultants enter the scene and attempt to help riparian parties to strengthen their internal capacity and to devise the legal and institutional frameworks to support these commissions. They aim to rewrite and standardize the national and international laws regulating water use and distribution between them. Often these international legal specialists lack specific in-country training, and instead rely on the body of international water law based on the Helsinki Rules and the International Law Commission recommendations to guide them through the web of bureaucracies in many of the countries they operate.[5] The international water lawyers who initially helped the Central Asian states have worked in countries as diverse as Sri Lanka, Albania, and Thailand.

In short, a certain kind of isomorphism in taking place across the Central Asian states in which the international community seeks to impose similar institutional structures throughout the region.[6] They want to create a similarity in form across the Central Asian states. Indeed, not only states (Scott 1998) but also international organizations prefer legibility. As mentioned, international water lawyers encourage states to adhere to the same body of water law and to standardize domestic water law. In the Fundamental Provisions of Water Management Strategy in the Aral basin, the Central Asian participants agreed that regional requirements must conform with international principles that include the right of each state in a basin to an equitable and reasonable share, sovereignty of each state over its natural resources, and the principle of no significant harm (ICAS 1996d, p. 17). International advisors exert influence by situating aid programs under the rubric of "sustainable development" in which many of these consultants and advisors then define the policy agendas for new states in the environmental field according to their notions of "sustainable development." In accordance with the requests of the donor community that the Central Asia states had overstressed the water management aspects of the Aral Sea problem and not the environmental aspects, they established the Sustainable Development Commission to handle the programs related to water supply, sanitation and health. The Central Asian leaders unabashedly expressed their willingness to join international conventions on the environment and to implement Agenda 21.[7] By professing their commitment to sustainable development, the Central Asian states received Capacity 21 funds from the UNDP to assist on developing a regional water convention, for example.[8]

USAID along with the EU-TACIS and UNDP, furthermore, promoted "water user associations" within Central Asia as part of numerous projects to encourage privatization within the agricultural and water sectors. This is also part of the sustainable development discourse in which international agencies seek to foster community participation (Daly 1996). In order to encourage internal self-financing, USAID adapted models from Mexico where grassroots water associations were successfully introduced. The goal was to relieve the state of the burden of upkeep and to take the costs of water use out of the state budget and to raise the price of water for the users.[9] With independence, the Central Asians have caught onto the importance of Western concepts and terminology, and as a result, many local water specialists often speak about the merits of these associations as an appropriate model for restructuring water use at the sub-national level. USAID, in response, held several training seminars at the national level in Uzbekistan pertaining to self-governing irrigation systems and on the development of a water pricing system covering irrigation, industrial, municipal, and domestic uses of water. Here, USAID consultants converged on the transfer of Western concepts concerning water markets, irrigation water user associations, and other pricing tools (Micklin 1997a).

Overall, the newly independent Central Asian states learned quickly that in order to attract international assistance they only needed to adopt the jargon of the Western IOs, bilateral aid organizations, and NGOs. They have become attuned to the art of role playing in the international system in which creating new international and domestic institutions is a form of compliance with the demands of the international community. Although in the water sector in Central Asia, consultants and representatives of IOs, bilateral aid organizations, and NGOs galvanize states to sign new agreements at the interstate level and try to shape the water and agricultural policies at the national level, much of this institutional restructuring has not led to real reform. Owing to side payments from the international community, President Karimov, for example, did not have to heed the demands of the opposition to stamp out cotton monoculture. Instead, Uzbekistan and Turkmenistan continued the system of goszakaz to procure cotton for foreign export.

In addition to creating a myth of new state institutions, the role of the international community has also contributed to the myth of nation building. On one level, the Central Asian states adopted traditional symbols of nationhood. They introduced new flags, national anthems, elevated the status of the titular language, declared new national holidays, and resurrected national heroes such as Tamerlane in Uzbekistan or the Manas epic poem in Kyrgyzstan from pre-Soviet times. On another level, they encouraged steps toward nation building that are not as evident to the visible eye. The water sector, particularly in Uzbekistan, afforded a subtle venue for nation building revolved around the perpetuation of a mythos of the irrigator and cotton cultivator (Pakhtakor).

As noted, during the Soviet period, most of the main scientific research institutes associated with water were located in the downstream state of Uzbekistan like the all-Central Asian Institute for Irrigation Research

(SANIIRI). The predominance of a cotton and water culture in Uzbekistan influences how the Central Asians themselves construct categories regarding one another. In various conversations with water specialists throughout Central Asia, the Uzbeks were always referred to as the "water people." In contrast, the Kazakhs were the "oil people," because of the rich petroleum deposits in the Caspian basin. These classifications have their genesis in the way in which the Soviet system of economic specialization designated individual republics or regions as responsible for one or another major economic input.

Since institutes like SANIIRI generated most of the information and data relating to the water system and irrigation use, researchers associated with the water sector in the other four basin states seldom question "Uzbek" authority on water issues. Uzbeks refer to the long history of sustainable irrigated agriculture in ancient Turkestan of which many of these early cites are located in present day Uzbekistan in the lower Amu Darya—Khorazm and Bukhoro, for example.[10] The references to pre-Soviet agriculture provide the Uzbekistan government and population with a myth of "Uzbek" agriculture as sustainable and environmentally sound. This myth suggests if Russian and Soviet planners had not intervened in Central Asia, the indigenous Uzbekistan population would have continued to work the land in a sustainable manner, maintaining their traditional patterns of rotation, and hence, averting the Aral Sea crisis.

Historical Uzbek classifications have also replaced the Russian water terminology, elevating the position of the mirab (the local water master). The Uzbekistan government has not tried to alter this myth of Uzbeks as water people and cotton farmers, but rather to reinforce it. As cotton still provided the backbone of the Uzbekistan economy, the government retained the ideal of cotton as "white gold." Finally, it has not hurt that the Western IOs have designated Uzbekistan as their base for operations. This empowered the Uzbekistan water sector to play a disproportionately large role to influence decisions in the water sector and to perpetuate the Uzbekistan government's faith in cotton cultivation and irrigated agriculture. Even when SANIIRI's influence began to decline at the national level, former leading scientists and engineers there guaranteed themselves a permanent position in the formal structure of the Aral Sea Basin Program through the creation of the Scientific Information Center.

Finally, the Central Asian leaders allowed for controlled NGO activity within their borders to reinforce nation and state building. NGOs were necessary to success as a "democratic" nation-state in the late twentieth century. In order to gain Western recognition, the Central Asian leaders had to show that they were upholding Western notions of a liberal democratic society. Thus, many of the Central Asian governments encouraged the birth of a small NGO community. With the creation of this highly censored NGO community, the Central Asian governments could claim to Western governments and aid organizations that they were truly engaged in political reform. However, the breadth of issues that local NGOs can address is limited by the political orientation of the regimes, which have all become more closed and repressive since independence. NGOs are often restricted to engaging in purely social and philanthropic activities and not in political activities. In Turkmenistan the government established its own human rights organization—Demokratii i Prav Cheloveka pri Prezidente Turkmenbashi (Democracy and Human Rights under President Turkmenbashi)—while repressing other indigenous organizations, such as the Russkoe Obshchina (Russian Society); the latter could threaten the government's control over society.[11]

In summary: The newly independent Central Asian states created new domestic institutions to conform to the demands of the international community in order to receive aid. They have adopted the discourse of democracy, markets, NGOs, sustainable development, along with water user associations. Although these institutions are situated within society, the Central Asian states have often failed in implementing them at the local level. Indeed, this process of institution building conforms to the notion of institutions as rituals wherein institutions are being constituted worldwide because of the myth of their usefulness for state building and nation building (Meyer and Rowan 1991).

Decoupled Institutions and the Soviet Legacy

As a derivative of this myth of usefulness, many of these new interstate institutions and even domestic institutions failed to produce their expected results at the national level. Although the international community stressed the need to establish a new institutional framework to address the Aral Sea crisis, many Central Asians first and foremost expressed their frustration that the rules governing water use and allocation had changed very little. Some local water experts suggested that only the international consultants and old apparatchiki were profiting from the international aid programs whereas many of them believed that they themselves were capable of devising their own solutions to their own problems.[12]

These new agreements are not really about determining water allocations and mitigating the Aral Sea crisis; rather, rapid regional cooperation among the Central Asian states was about maintaining the system of social and political control under conditions of transformation. The Central Asian leaders crafted these institutions because they were useful for procuring foreign aid. However, it turns out that in the short-term, they could not produce the expected result—ameliorating the Aral Sea crisis. In many instances, the Aral Sea Basin Program had not reached the farm level. This was the case because the Central Asian leaders constituted these institutions to create a perception of normalcy, primarily for the international community. The leaders engaged in interstate cooperation as a means of becoming members in the international community of nation-states. Since it was understood that institution building served as the condition for securing aid, the Central Asian leaders grafted these new institutions onto pre-existing institutional arrangements. The end result was that these new institutions were decoupled from what they were supposed to do locally.

Moreover, even if an agreement was signed, local actors could easily overlook them, even blatantly disregard them. Despite a protocol signed between the Uzbekistan government and USAID on providing technical assistance to the government of Uzbekistan in the area of managing water resources, the USAID resident advisor—the geographer Philip Micklin—found many obstacles in his path before he could formally initiate this project. While the Uzbekistan government exalted the discourse of water reform and domestic water associations, the government's First Deputy Minister of Reclamation and Water Management (later appointed Deputy Minister of Agriculture and Water Management), Abdurahim Jalolov succeeded to stymie the protocol's implementation. He refused to provide an office for the resident advisor and directly expressed his lack of interest to work with the resident advisor, claiming that Uzbekistan had enough local know-how and expertise in water management (Micklin 1997a).

Although outwardly committed to the World Bank, European Union-WARMAP, and USAID programs, the Central Asian leaders also clung to past methods of "resolving" their problems, such as negotiating separate barter agreements over water and fuel resources. In 1995 and 1996, while working within the new framework for water management, the Central Asian states, nevertheless, produced a series of bilateral and trilateral agreements to correct water allocations made along the old Soviet schemes for water distribution. Whereas the Central Asian leaders had devised a new framework and organizational bureaucracy for water management in order to meet the international community's conditions for aid, the influence of the Soviet legacy during this time continued to undergird these new institutional arrangements. In turn the World Bank and the EU-WARMAP adjusted their projects to incorporate these past Soviet structures for water management into new ones. USAID went a step further and constituted new negotiations—the Water and Energy Uses Roundtable—that would directly link energy with water.

While the domestic context remained precarious with independence shifting the balance of interests and capabilities among the states, the immediate intervention of the international community helped to stabilize the domestic political and social climate. It also served to bolster vested interests in the water sector. As a result, the international community ended up cushioning these post-communist states from undergoing fundamental economic and political change. This enabled the Central Asians to keep doing things as they had always done while presenting the illusion of change through new institutions to the international community. International intervention brought continuity to a situation in which the rules of the game were in a state of flux. The act of signing agreements in the water sector and other interstate agreements related to customs, citizenship, and borders squelched the shared fears of what the transition could potentially hold.

International intervention, therefore, did not mean that the institutions constituted were the most efficient or best suited for rectifying the Aral Sea crisis. Indeed, the shared fears among the domestic actors and the international community that conflict could potentially erupt between independent states allowed for the legacy of the Soviet system of water management to infringe on any new changes in the formal system of water management. Without a tradition of rule of law, the Central Asian states affirmed their commitment to the principles of international law in public statements and declarations, while domestically past patterns of prior use and local customs continued to underlie practices governing water use and allocation. In short, the Central Asian states codified these new institutional agreements onto pre-existing Soviet social structures. For example, the new apex agreements did not replace the role of the ICWC composed of the five former water ministers. Instead, the ICWC remained an integral part of the new Aral Sea Basin Program, retaining the actual authority over water-allocation decisions. ICAS (later IFAS) in the end was only supposed to make the decisions that the ICWC could not assume.

Even if the collapse of the Soviet Union provided a real opportunity to rectify past policies and institutional mechanisms for water management, the Central Asian water institutions tended to be quite conservative and resistant to change as fundamental change was a process with high fixed costs. Although the Central Asian leaders devised new institutions and signed new agreements, they have to a large extent been path dependent and highly contingent on the previous institutional structure that was incredibly inefficient and socially and environmentally deleterious. In essence, the system of centralized water management and cotton monoculture offered a mechanism for both political and social control.

International institutions remained decoupled from local patterns of water management because international actors initially focused on the establishment of new interstate agreements and organizations to deal largely with the potential for water conflicts. They did not direct their efforts toward connecting the Aral Sea Basin Program to the farm level during the first stage of active intervention. When interviewing the head of Oshvodkhoz (Osh Water Authority) in Osh, Kyrgyzstan in 1995, it was clear that he had not been informed of the new interstate water-sharing framework, as he appeared puzzled by the mention of it.[13] Other employees remarked that most questions of water use and distribution were still decided at the local level. They eagerly provided examples of cooperation among Osh (Kyrgyzstan), Andijon (Uzbekistan), Fergana (Uzbekistan), and Leninobad/Khujand (Tajikistan) at the oblast level that had nothing to do with the interstate agreements, but rather with merely picking up the phone. Moreover, they emphasized that it would be impossible for them to follow any standard principles of international law because they could contradict Islamic law, which "forbids the charging for water under most circumstances."[14] Clearly, those engaged in the creation of the new interstate bodies were detached from the local level by excluding the local water managers in the process of creating new water institutions. To prevent international institutions from remaining decoupled from local practices, it is necessary to expand the role for local actors and community participation in the various action plans. At the very least, the linkages among the villages, capital cities, and the international community needed to be better clarified. This is essential so the farmers see a direct link to the changes taking place at the interstate level.

These new states acted in accordance with international norms of statehood in order to become viable states and to ensure a continuous flow of financial and material resources from abroad. They adhered to the advice of international actors and built new institutions, engaged in forms of interstate cooperation, and signed onto new agreements. Yet many of these water institutions were inefficient, especially in the area of environmental protection. Simply establishing a collection of institutional arrangements provided no assurance that the function of environmental management would be fulfilled. Most of the international community's efforts concentrated on strengthening the capacity of these new interstate water institutions, but in order to support these measures, institutional change was also necessary at the national and local levels. Thus, in the long-term, it will not matter if new interstate allocation and distribution rules are in place if they cannot be enforced at the national and local levels.

In summary: The international community had the power to force states to adopt new domestic institutional structures, but at the same time, this did not mean that these institutions would function in the positive way that the international community expected. Instead, in Central Asia we find many perverse effects in which the new domestic institutions are disjointed from the empirical reality.

Too Much International Intervention

Another implication of this enlarged role for the international community is that third-party actors began to perform the traditional role of states by bolstering previous domestic institutions and vested interests. Instead of state elites responding to the concerns of their domestic constituencies, IOs, bilateral aid organizations, and NGOs assisted those groups hardest hit by the transitions. Transnational actors negotiated with both state elites and local actors, and as a result, both state elites and local actors put off redefining the terms of new state-society relations after independence.

Both state elites and local actors hoped the international community would provide a quick solution to cushion the shock of the transition away from state socialism. Consequently, many advisors and consultants feared that "donor community may, inadvertently, be developing a 'international welfare mentality' among the aid recipients in the Aral basin" (Micklin 1998, p. 11). One representative from TACIS commented that a situation exists in which "professional givers are creating a nation of professional takers."[15] Rather than encouraging the Central Asians to derive their own solutions to Aral Sea crisis, the international community may have jumped in too early if it sought to mitigate the environmental degradation caused by cotton monoculture. The international community replaced Moscow as a supplier of resources without being able to meet its objective of environmental protection in addition to regional cooperation.

At times, the policy of inclusion was extremely problematic. On the one hand, outside actors needed to work with the old guard who were responsible for the Aral Sea problem. On the other, if outside actors overlooked the old guard, the program might have been subverted. As a result, the World Bank and donors such as the EU-TACIS collaborated with the five-country ICAS and its related organizations to develop mechanisms for water allocation and management in the Aral basin. In contrast, USAID concentrated less on the old water nomenklatura associated with ICAS and ICWC, but chose instead to work with the ICKKU.

Multiple negotiations under the auspices of several competing donor groups resulted in a number of draft agreements and too many interstate organizations claiming jurisdiction over water management. In contrast, lessons from other regional attempts at environmental cooperation such as the Long-Range Treaty on Air Pollution show that even if an agreement is weak and vague at the outset, subsequent protocols can strengthen it (Levy 1993). Whereas duplication and redundancy in many of the Aral basin programs was common as a result of the overzealousness of the international community to see immediate results. Moreover, each third-party actor sought to claim "success" in bringing about interstate cooperation. This multi-dimensional approach forced local actors to compete among themselves for access to international resources; at the same time, it also enabled local groups to play off outside actors against one another. Each international organization was competing to be the first to assist the Central Asians in reaching an interstate agreement. Nevertheless, their consultants frequently complained about the slowness of the pace while ironically most major environmental agreements have taken extended periods of time before even a weak convention was achieved. For instance, it took 10 years to produce the final text of the UNCLOS (UN Convention on the Law of the Sea). Even after more than 25 years of cooperative efforts, the Mediterranean basin states are still only working on the second phase of environmental cooperation, which is characterized by implementation efforts and building national capacity to ensure compliance (VanDeveer 2000). In 1998 a cooperative endeavor sponsored by UN agencies and the World Bank led to the creation of the Caspian Environment Programme.[16] Yet these efforts have failed to address the issue of legal demarcation, but rather have focused on pollution prevention, monitoring, biodiversity protection, and management of sea-level fluctuation.

The Formative Years of State Building: Getting It Right or Getting It Wrong

The thrust of this book has demonstrated that side payments formed the crucial link between the international and the domestic level in which third-party actors promoted the use of side payments to assist the dual institution-building process. Although third-party actors succeeded in fostering regional environmental cooperation, their immediate intervention raises the issue of the long-term effects of having IOs, bilateral aid organizations, and NGOs play such a tremendous and influential role in the formative years of post-communist state building. IOs, bilateral aid organizations, and NGOs intervened immediately in Central Asia to prevent conflict from erupting over scarce water resources, especially in view of the history of resource conflicts in the region. Since the Central Asian successor states were poor and weakly institutionalized at independence, the early role played by IOs, bilateral aid organizations, and NGOs enabled them to have a lingering impact on the future trajectories of political and economic development in the region and, more important, on state-society relations.

The process of state making under conditions of transformation concurrently afforded the Central Asian leaders newfound opportunities to alter both internal state-society relations within them and external relations among them. During this transition period, the Central Asian states could have broken with the past Soviet institutions or chosen to reinforce them even though the structure of the international system had changed. Internally, the Central Asian leaders faced a strategic choice whether to maintain Moscow's policy of cotton monoculture or to diversify agriculture; the latter entailed the creation of independent farms in place of collective ones. In the context of state formation, diversifying agriculture required dismantling the system of social control in which farmers received subsidized agricultural inputs and other social goods in return for political loyalty. In order to dismember the system of social control, the Central Asian leaders needed to reconstruct the relationship between state and society in which loyalty would no longer be dependent on patronage but would have to be based on a reciprocal exchange linked to taxation and representation that is characteristic of the modern Weberian state.[17] However as chapter 7 demonstrated, the Central Asian leaders resisted such changes in the short-term because they preferred to retain Soviet methods of internal rule.

Although the aid channeled to the Aral Sea Basin Program was on a much smaller scale than what Russia received from the IMF or the World Bank to foster markets and democracy, it, nevertheless, cushioned the deteriorating economic and social situation in the Aral basin. It, furthermore, enabled the leaders to delay radical economic and political reform similar to that, which had been proscribed for Eastern Europe and Russia after the collapse of the Soviet Union. In particular, the Central Asian leaders solicited side payments from the international community to not only address the Aral Sea crisis, but more important to pay off vested domestic interests that could challenge the leaders' authority and rule. The Central Asian leaders, in turn, channeled these side payments from the international community to important domestic interests, enabling the elites to strengthen domestic patronage systems within the context of independent states. Although the acknowledged goal was to ameliorate the Aral Sea crisis and to foster democratic and market institutions, side payments in the end helped the Central Asian states resist domestic reform related to the cotton sector right after independence. By putting off reform early on in a transitional period, it was difficult for the Central Asian leaders later to introduce sweeping reform that could cause economic hardship. Rather than encouraging the Central Asian states to "swallow the bitter pill," the international community caused the Central Asian states to miss the crucial period in a transition when a population may be responsive to radical change.[18]

In fact, because stability became the preferred option for Western policy makers and Central Asian leaders alike, the international community was unwilling to push the Central Asian states to follow through on structural adjustment and privatization programs like those which took place in East Central Europe.[19] States with strong opposition movements usually fail to introduce reforms because it is difficult to keep economic reformers insulated from politics and the pressure from vested interests, such as labor unions. Some argue that the first stages of the economic transformation in East Central Europe indeed benefited from the disarray and lack of organization among the major groups opposed to market-oriented reforms (Nelson 1993). Ironically, in the Central Asian states where the opposition had been most suppressed—Turkmenistan and Uzbekistan—the leaders failed to adhere to the advice of the World Bank and the IMF to reform their economies. Because the international community did not want to challenge stability in the region, it, in turn, did not exert extreme pressure on the Central Asian states to undertake sweeping economic or democratic reform. In fact, when Kazakhstan disbanded its parliament in March 1995, the international community stood by silently.[20] The downside of this preference for political stability over economic and political reform is that members of society and former members of the opposition have lost faith in the international community as real proponents of change.

Lastly and most important, owing to the flow of resources from the international community from the sale of export commodities, the new governments were able to delay introducing new taxation systems. By continuing to rely on cotton procurements, the Uzbekistan and Turkmenistan governments did not have to introduce a system of taxation since the cotton monoculture was an indirect form of taxation on the population. Cotton revenues filled the state coffers rather than domestic taxation. The result was that the reciprocal relations necessary for democratic institutions to take root were not emerging. The notion of an exchange between taxation and representation never entered the political discourse during the early stages of Central Asian state formation. Likewise, in Kazakhstan oil and gas rents have delayed the creation of a viable tax regime (Weinthal and Jones Luong 2000). Thus, similar to other states dependent on either foreign aid or foreign revenue from a single commodity, the governments were freed from levying domestic taxes, and in turn they did not have to be accountable to their domestic populations and to build the reciprocal linkages between state and society.[21]

Conclusion

By highlighting the formative years of state making, this book illuminates the power of IOs, bilateral aid organizations, and NGOs between 1992 and 1998. An enlarged and proactive role for IOs, bilateral aid organizations, and NGOs in the post-Cold War context helped prevent interstate conflict over fresh-water resources. This in of itself is a remarkable achievement, especially since the amount of aid was on a much smaller scale than in other post-communist states. Without aid recurrent conflicts like those that took place at the end of glasnost over scarce resources or that which erupted between Hungary and Slovakia over the construction of the Gabcikovo-Nagymaros dams after the breakup of Czechoslovakia might have occurred. Yet at the same time, the Central Asian leaders and the international community did not immediately resolve their environmental problems. Rather, the enlarged role for transnational actors inadvertently propped up these weak states and prevented the introduction of sweeping economic and political reform. The Central Asian leaders professed their commitments to democratic and market reform while continuing to become more repressive toward their domestic populations. Therefore, the challenge still remains for IOs, bilateral aid organizations, and NGOs to ensure that their promises of aid to the Central Asian states result in building the necessary domestic capacity to finally tackle the environmental legacy associated with cotton monoculture.

Owing to the dire need for aid, third-party actors began to usurp the role that Moscow once played as the dispenser of patronage, and, furthermore, influence the internal components of these new states. This mimicked earlier political developments in Africa wherein international actors through their colonial and then post-colonial policies molded the relationship between state and society (Migdal 1988; Rothchild and Chazan 1988). As in Africa decades earlier, international intervention propped up weak states in Central Asia. Yet the scope and form of international intervention has differed significantly in the Central Asian context owing to the use of environmental aid and to the nature of the post-Cold War context. Whereas during the Cold War era, the superpowers provided aid to African states to help them fight proxy wars that corresponded to the East-West divide; in the post-Cold War context states are not the main suppliers of aid, but rather non-state actors such as IOs and NGOs. Within this changed political landscape, third-party actors have shied away from direct or indirect military intervention to more humanitarian goals, which include addressing environmental and health problems. Yet the Central Asian case shows that even small-scale environmental aid can also have unintended consequences for political and economic development. This effect is more pronounced with the Soviet Union's collapse because without an alternative ideological system available, the post-communist states could only turn to Western aid organizations to help them weather the transitions away from state socialism. In short, IOs, bilateral aid organizations, and Western NGOs have much more power available to them in the post-Cold War context than during previous periods. While much attention has been paid to their positive attributes at the interstate level in which they help resolve collective-action problems and contribute to finding solutions for global environmental problems, less emphasis has been placed on the negative effects of IOs on the internal politics of states. This book has sought to address this crucial aspect regarding the negative consequences of transnational activities domestically despite the positive aspects at the interstate level.

In general, the Central Asian case points to the changing role of the international community in the post-Cold War context in which states are no longer the main suppliers of technical and financial assistance, but rather non-state actors. This, in turn, generates specific lessons for other states in the making in the twenty-first century. Lessons from Central Asia are highly relevant for the state formation process within the Palestinian territories. In particular, the signing of the Declaration of Principles of Interim Self-Government Agreements (known as the Oslo Accords) ushered in a burgeoning opportunity for state and non-state actors in the Middle East to deal with issues of water scarcity and environmental degradation. Similar to the Central Asian states, the Palestinian Authority must simultaneously conclude a water-sharing agreement over the West Bank aquifers with Israel while also building the empirical institutions of statehood. As in the Aral basin, IOs, bilateral aid organizations, and NGOs are assuming an active role in supporting both regional cooperation and domestic state building. Part of the rationale for being in the Palestinian Authority is not to provide "normal aid" but rather to render "political aid" through economic and environmental assistance programs.[22] Here too, aid is not just directed at supporting regional cooperation and environmental protection in the water sector, but more so, it is tied to building robust states that can enter the international community of nation-states.

For newly emerging states, political processes at both the domestic and international levels will undoubtedly be embedded within a complex and varied network of transnational relations among state and non-state actors. As the activities of these actors become more pronounced, it is essential to pay attention to the way in which their intervention and side payments may affect the state-building process in ways not foreseen. Despite many of the practical failures of implementation, international efforts to address the Aral Sea crisis illustrate how states are constructed within the context of the international system and furthermore how national elites use institutional symbols for domestic legitimacy.

Appendix
The Aral Sea Basin Program

(source: briefing paper for proposed meeting of donors, World Bank, 1994)

Program 1

1. Regional Water Resources Management
2. Improving Efficiency and Operation of Dams
3. Sustainability of Dams and Reservoirs

Program 2

1. Hydrometeorological Services
2. Database and Management Information System for Water Quality and the Environment

Program 3

1. Water Quality Management
   1st)Water Quality Assessment and Management
   2nd) Agricultural Water Quality
2. Collector Drains

Program 4

1. Wetland Restoration
2. Restoration of Northern Part of the Aral Sea
3. Environmental Studies in the Aral Sea Basin

Program 5

1. Clean Water, Sanitation and Health—Uzbekistan (short-term)
2. Clean Water, Sanitation and Health—Turkmenistan (short-term)
3. Clean Water, Sanitation and Health—Kazakhstan (short-term)

Program 6

1. Integrated Land and Water Management in the Upper Watersheds

Program 7

1. Automatic Control Systems and Civil Works for the Amu Darya Basin, including Capacity Building for BVO Amu Darya
2. Automatic Control Systems and Civil Works for the Syr Darya Basin, including Capacity Building for BVO Syr Darya

Program 8 (Supplementary Program)

1. Capacity Building for EC-ICAS and IFAS

Notes

Chapter 1

1. Quoted on p. 76 of Ellis 1990.

2. Kamalov 1996, p. 25.

3. By 1991, the salinity had reached 37 grams per liter. For a comprehensive overview of the Aral Sea tragedy see Glantz 1999.

4. Source: *Aral: Yesterday and Today* (International Fund to Save the Aral Sea, UNDP, and World Bank, 1997), pp. 23–24. The disappearance of 20 of the 24 native fish species is attributed to the rise in salinity.

5. According to Micklin (1991, p. 4), the estimated average annual river flow in Central Asia is 122 km^3. The Aral Sea drainage basin accounts for 90% of this, of which the Amu Darya's annual flow is 73 km^3 and the Syr Darya's is 37 km^3.

6. An oblast' is a political subdivision comparable to a province. Hereafter, for typographical reasons, the apostrophe will be omitted from oblast', glasnost', and similar words.

7. See also T. Kaipbergenov, "Tri uroka Arala," *Pravda vostoka*, January 28, 1990.

8. Bohr (1989, p. 37) reports 111 deaths per 1000 live births in Karakalpakstan during this period.

9. I observed this during visits to Nukus and Muynak, Uzbekistan, in 1994.

10. Kaipbergenov, "Tri uroka Arala."

11. Afghanistan is also an important riparian state in the Amu Darya basin.

12. *Kyrgyz* can be singular or plural. *Kyrgyzstani* refers to a citizen of Kyrgyzstan, whether a Russian, a Kyrgyz, or an Uzbek; *Kyrgyz* refers only to an ethnic group.

13. *Pravda vostoka*, December 22, 1992, quoted on p. 150 of Critchlow 1995.

14. Side payments are extra payoffs (not parts of the agreement itself) that are used to get actors to reach an agreement. They may be paid to the actors sitting

at the table, or they may be paid to actors not at the table who have the potential to weaken or to interfere with the agreement made there.

15. Others, including Anderson (1999) and Moore (1998), have focused on many of the moral problems associated with rendering aid.

16. *Nomenklatura* refers to those who occupied the most prestigious positions in the Communist Party apparatus.

17. See e.g. Homer-Dixon and Blitt 1998; Baechler and Spillman 1996; Hauge and Ellingsen 1998; Esty et al. 1995.

18. On the importance of institutional design, see Keohane and Ostrom 1995; Mitchell 1994; Victor, Raustiala, and Skolnikoff 1998; Weiss and Jacobson 1998.

19. For a similar interpretation see p. 16 of Wedel 1998.

20. For an example see Maoz 1996.

Chapter 2

1. For an overview see Dabelko and Dabelko 1995.

2. For a skeptical view see Deudney 1990.

3. Yet many scholars now recognize that the development of state capacity helps to preclude such events from spiraling into conflict-prone situations.

4. On attempts to tease out these linkages, see the case studies in Homer-Dixon and Blitt 1998. See also issue 2 of *Environmental Change and Security Project Report* (Woodrow Wilson Center, spring 1996), especially the comments by Jack Goldstone.

5. Europe has four river basins that are shared by four or more countries, but conflicts are rare because these rivers are regulated by approximately 175 treaties (Clarke 1993, p. 92).

6. On conflict and cooperation over water see Gleick 1993a; Barrett 1994; Frey 1993; Crow et al. 1995; Elhance 1999; Lowi 1993a; LeMarquand 1977; Ohlsson 1995.

7. This definition covers river and lake navigation and issues pertaining to use, development, and conservation in surface, underground, atmospheric, and frozen waters.

8. Updated numbers: Danube (17 after the breakup of Yugoslavia, Czechoslovakia, and the Soviet Union), Niger (10), Nile (10 with the inclusion of Eritrea), Zaire (9), Rhine (8), Zambezi (8), Amazon (7), Mekong (6), Lake Chad (6), Volta (6), Ganges-Brahmaputra (5), Elbe (5), La Plata (5). The Aral Sea Basin should also be included in this list, since it is shared by the five Central Asian states along with Afghanistan (6).

9. Similarly, groundwater systems create different incentives for cooperation, depending on how the water flows. For example, the recharge area and the early flow of the mountain aquifer shared by Israel and the Palestinian Authority are in Palestinian territory, whereas the downflow is within the state of Israel.

10. Oran Young (1994a, pp. 19–26) offers a useful typology for simplifying the range of environmental problems, but much overlap among the categories still remains. Young classifies environmental problem sets accordingly: international commons, linked issues, shared natural resources, and transboundary externalities.

11. See e.g. Barkin and Shambaugh 1999.

12. See also Ostrom, Gardner, and Walker 1994.

13. Ostrom (1990) lays out three central puzzles associated with collective action problems in common-pool resource systems: the problem of supply, the problem of credible commitments, and the problem of mutual monitoring.

14. For ways in which heterogeneity is operationalized, see the special issue of *Journal of Theoretical Politics* 6 (1994) on local commons and global interdependence.

15. These positions mirror what international water law refers to as *customary law*. For an overview of customary water law and subsequent attempts to codify alternative principles of water use based upon equitable utilization and an obligation that states should not cause harm to others through their development programs and water use, see McCaffrey 1993 (pp. 92–104) and Caponera 1985.

16. For an overview of the conflict see Lipschutz 1998; Klötzli 1993; Fitzmaurice 1996.

17. According to World Resources Institute et al. 1996 (pp. 301–302), per-capita water availability is measured as the annual renewable water resources per capita that are available for agriculture, industry, and domestic use. See also Gleick 1993a.

18. In the period 1991–1993, according to the Food and Agriculture Organization of the United Nations, 91% of Turkmenistan's and 92% of Uzbekistan's cropland was irrigated. Even in Kyrgyzstan, a water-rich country, 67% of the cropland was irrigated. Source: World Resources Institute et al. 1996, p. 241.

19. This is analogous to Waterbury's (1994) classification of patterns of cooperation in the water basins of the Middle East.

20. See e.g. Keohane 1984; Krasner 1976.

21. For an overview of the policy implications see chapter 1 of Ostrom 1990.

22. For examples see Matthew 1999, especially pp. 164–168.

23. On differences between bargaining leverage and structural power see pp. 117–139 of Young 1994a.

24. For an overview see Hardin 1968.

25. Taylor and Singleton (1993) find that when a community exists where individuals share similar beliefs and values, individuals are more likely to interact with and trust one another precisely because relations are direct, many-sided, and reciprocal. See also Taylor 1987.

26. Khazanov (1992) describes the governance practices for water management in the pre-Soviet period as decentralized.

Chapter 3

1. I recognize that "IO" can also serve as an umbrella grouping for both intergovernmental organizations and international non-governmental organizations, but in order to be parsimonious with acronyms I use "IO" to refer to intergovernmental organizations. I then use "NGOs" for non-governmental organizations, knowing that they may be indigenous or international. I am looking at actors that usually have transnational activities attached to them. Since this is a study of transnational actors, I also look at the role of bilateral development agencies as actors in their own right. For an early work on IOs see Jacobson 1984.

2. On the role of IOs, bilateral aid organizations, and NGOs in Central Europe, see Wedel 1998. See also Mendelson and Glenn 2000.

3. Similarly, each state in the Caspian basin sought to define its own energy-development strategy. See Jones Luong and Weinthal 2001.

4. On the principle of "permanent sovereignty over natural resources," see p. 113 of Birnie and Boyle 1993. The UN General Assembly adopted this principle in 1962 with resolution 1803 XVII.

5. Several scholars have focused on the ways in which the Central Asian states have pursued different paths of state formation. See e.g. Olcott 1996 and Gleason 1997.

6. At the international level, cooperation requires joint activity among two or more states. According to Keohane (1984, pp. 51–52), cooperation arises "when actors adjust their behavior to the actual or anticipated preferences of others, through a process of policy coordination." Moreover, by focusing on the development of new institutions for water sharing in Central Asia, the case of the Aral basin helps us to untangle the overarching discrepancies between form and content that often plagues the study of institutions. Observing how the process of institutional design unravels will allow us to better understand when institutions actually contribute to mutual policy adjustment or whether they are just merely epiphenomenal. Clearly, it is possible to have policy adjustment without institutions and likewise institutions without policy adjustment. However, I am concerned about the causal relationship in which institutions are able to encourage policy adjustment.

7. According to Oran Young, the concept of institutions is distinguishable from international organizations and not interchangeable. Young (1994a, pp. 3–4) defines institutions as "sets of rules of the game or codes of conduct that serve to define social practices, assign roles to the participants in these practices, and guide the interactions among occupants of these roles," whereas organizations are the "material entities possessing offices, personnel, budgets, equipment, and more often than not, legal personality."

8. For a detailed discussion of collective action problems that exist "where rational individual action can lead to a strictly Pareto-inferior outcome, that is, an outcome which is strictly less preferred by every individual than at least one other outcome," see p. 19 of Taylor 1987.

9. Haggard and Simmons (1987, p. 495) derive this definition from Oran Young's work on resource regimes. They also point out that this definition "allows a sharper distinction between the concept of regime and several cognates, such as cooperation."

10. On the specific role of counterfactuals see pp. 3–38 of Tetlock and Belkin 1996.

11. This is not to say that Central Asia was immune from conflict, but many of the small-scale ethnic conflicts related to water and land resources that took place in the late 1980s were absent. Rather, conflicts were regional in nature like the civil war in Tajikistan. Other conflicts that took place include the shootings in the Fergana Valley during 1999.

12. The exception here is Belarus, which for all intents and purposes has chosen reintegration with Russia. On the failure of the Commonwealth of Independent States to materialize, see chapter 3 of Olcott 1996.

13. For an overview of Caspian politics see Ebel and Menon 2000.

14. Relevant works include Axelrod 1984, Keohane 1986b, and Oye 1986.

15. But cooperation among the advanced industrialized countries should not be overstated because even in these cases, the path to cooperative agreements still entails a long and drawn out process. For example, the Long-Range Convention began with a framework convention followed by a series of protocols to clarify the regime.

16. On the effects of time on cooperation see Matthews 1996. For a general overview of the debate on relative gains see Grieco 1993.

17. Even transitologists and consolidologists recognize the uniqueness of these transitions. See e.g. Schmitter and Karl 1994.

18. On previous periods of state formation see Tilly 1975, 1990; Giddens 1987; Rubin 1995; Jackson and Rosberg 1982.

19. Neorealism or structural realism parsimoniously systematized the basic tenets of realism that arose as a consequence of the inter-state conflict early in the 20th century. See Waltz 1979 and Keohane 1986a. In this chapter, I use these terms interchangeably. For a collection of articles concerning the differences between the two approaches see Baldwin 1993.

20. One exception is Zeev Maoz (1996).

21. Schelling (1960) makes this observation.

22. See e.g. Schreurs and Economy 1997, especially the chapter by Robert Darst ("The internationalization of environmental protection in the USSR and its successor states").

23. For a similar conclusion on Western Europe see Moravcsik 1994.

24. Some representative works: Mendelson and Glenn 2000; Risse-Kappen 1995; Wapner 1995; Princen and Finger 1994; Clark, Friedman, and Hochstetler 1998; Barnett and Finnemore 1999; Keck and Sikkink 1998. Two early works

that looked at transnational actors: Keohane and Nye 1972 and Huntington 1973.

25. Concerning various views and arguments on the types of international intervention in domestic affairs, see Helman and Ratner 1992–93; Lyons and Mastanduno 1995.

26. For a similar case of regional cooperation among transitional states in East Central Europe, see Bunce 1997.

27. During the years immediately after the collapse of communism in the East, the newly independent post-communist states embraced the language of democracy and markets as they saw little other alternative if they wanted financial and material assistance to weather the transition. A large literature on these democratic and market transitions ensued. For a few examples see Przeworski 1991; Clague and Rausser 1992; Colton and Legvold 1992; Boycko, Shleifer, and Vishny 1997.

28. See e.g. Princen and Finger 1994; Keck and Sikkink 1998; Keohane and Levy 1996.

29. In the introduction to Evans et al. 1993, Andrew Moravcsik begins to adjust Putnam's model to include other actors such as transnational alliances. Barnett and Finnemore provide an example of a more nuanced study of IOs as actors with interests. See also Barnett and Finnemore 1999.

30. Much of this confusion stems from early research within the subfield of international organization that focused mainly on descriptive studies of the legal and constitutional aspects of the League of Nations and later the United Nations. According to this conventional literature on international organizations, institutions were viewed synonymously with formal organizations. Scholars did not treat these organizations as independent actors, but rather saw them as dependent variables, reflecting a movement toward integration and the decline of the nation-state (Rochester 1986).

31. See e.g. p. 702 of Conca 1994.

32. On functionalist or neofunctionalist arguments, see e.g. Mitrany 1966 and Haas 1964.

33. Side payments are forms of compensation to induce an agreement or cooperation among actors. Again see Schelling 1960, p. 31.

34. Risse-Kappen (1995) also highlights the importance of domestic structures in determining the nature of transnational relations or the way in which domestic structures interact with non-state actors.

35. On the dynamics of the state socialist economy see Kornai 1992.

36. Whereas I view the Soviet Union as a federal state, others are inclined to place it within the framework of empire. For an overview of this debate see Suny 1995. On the breakup of the Soviet Union, Yugoslavia, and Czechoslovakia, see Bunce 1999.

37. On quasi-states see Jackson 1990.

38. On the role of UNEP see Tolba with Rummel-Bulska 1998.

39. See also Keck and Sikkink 1998, p. 2.

40. This is similar to what Robert Jackson (1993) describes as the transfer of ideas. Here notions of sovereignty and independence constitute the feasible set of alternatives for the successor states of the Soviet Union.

41. For example, in Central Asia—see Jones Luong and Weinthal 1999.

42. For an anthropological critique of the developmental model propounded by the World Bank and of the discourse associated with development, see Ferguson 1994.

43. This is similar to Ostrom's (1994) discussion of the head-end and tail-end problem in a local irrigation system.

44. For an early assessment of the GEF see Fairman 1996.

45. Initially, the European Bank for Reconstruction and Development set out to help facilitate political reform as well as economic reform in the East Central European countries. This goal differed remarkably from how other banks operate as simply lending institutions. See Stein 1996.

Chapter 4

1. For an explanation of this term see Jones Luong 2000. See also Thelen and Steinmo 1992.

2. For a historical overview of the conquest of Central Asia see Allworth 1994. After capturing Toshkent, Russia turned its attention toward the three khanates that compromised most of Turkestan, beginning with the Bukhoran Emirate. After the defeat of Bukhoro in 1868, the Khanate of Khiva fell in 1873. The Kokand Khanate was fully subdued in 1876. On Russia's presence and policies in Turkestan see Pierce 1964.

3. For a longer-term perspective on the interests of Russian cotton manufacturers in relation to other industrialists in Czarist Russia, see Joffe 1984. See also Lipovsky 1995.

4. In actuality the mountain chains are a third geographic region, but besides a small percentage of pastoralists there most inhabitants lived in either the steppes or the oases. Many Western observers have assimilated these delineations into their own studies of Central Asian culture and history. See e.g. Bacon 1966. Ironically, in the post-Soviet period, many historians have further sought to ossify such cleavages between nomads and settled people to bolster a new national identity for the Soviet successor states. See e.g. Masanov 1995.

5. Although formally incorporated into the Russian Empire, the Bukhoran and Khivan khanates were allowed to retain their autonomous status until the Russian Revolution of 1917. Bukhoro, in particular, became a state under Russian suzerainty.

6. Other source: interview with S. Mirzaev, Rector of Toshkent Institute of Engineering for Irrigation and Agricultural Mechanization, January 20, 1995.

7. For a discussion of early irrigation systems see Sirozhidnikov 1991.

8. Although much of the agriculture was based on dry farming, I focus on the oases where irrigation was used. On irrigation and agricultural patterns in the Middle Zarafshon Valley at the end of the 19th century and beginning of the 20th century, see pp. 50–74 of Rassudova 1969.

9. For an example of traditional methods see Karimov 1998.

10. Graf K. K. Palen toured Turkestan in 1908–09 in order to prepare a report on a number of recommendations to the Czarist government for reform in Turkestan. One area he focused on was irrigation. See Palen 1910.

11. Again, the appointment of a local water master is consistent with water administrative practices throughout Muslim societies. On how the role and the duties of the water master have varied in Muslim societies, see p. 73 of Caponera 1992.

12. For examples see p. 647 of Rassudova 1969 and p. 280 of Matley 1994. Later the governor-general tried to exercise authority over the appointment of the ariq aqsaqal (Palen 1910; Abramov 1916).

13. For a more nuanced version of the Soviet delineation of Central Asian beyond just "divide and rule," see Hirsch 2000.

14. This decree became the objective of a subsequent decree (May 17, 1919) on the importance of irrigation for cotton independence (Dukhovny and Razakov 1988, p. 27)

15. "The Collectivization Campaign in Uzbekistan," *Central Asian Review* 12 (1964): 40–52. On the particular details of the land and water reforms, see Igamberdiyev and Abdurakhmanova 1975; Aminova 1974.

16. "The Collectivization Campaign in Uzbekistan," p. 42.

17. For a description of different patterns of social organization among nomads see Winner 1963a. The uru was a larger form of social organization; several uru formed a tribe. When summer came, the uru would divide up for herding migrations.

18. See also *Irrigatsiia Uzbekistana,* volume 1 (FAN, 1975).

19. "The Hungry Steppe," *Central Asian Review* 5 (1957): 42–48.

20. For an overview of earlier attempts at resuming work in the Golodnaya Steppe during the Russian migration into Turkestan, see pp. 152–154 of Bartold 1927.

21. As on other farms that were consolidated from previously existing villages, a whole Korean community would settle one farm, enabling it to develop strong patronage networks. A few of these farms were located in the Golodnaya Steppe, but the most infamous of the Korean run farms is Politodel Kolkhoz in Toshkent Province. On-site investigation, Politodel, May 1992. For a general overview of the role of the Korean community in agriculture in Central Asia, see Shim 1995.

22. Irrigated Crop Production Systems, volume IV, TACIS, Water Resources Management and Agricultural Production in the Central Asian Republics, January 1996, p. 2.

23. Estimate from Irrigated Crop Production Systems, volume IV, p. 1. For comparative purposes, this is the same figure generally thrown out to describe water use in California.

24. Gleason (1990a, p. 67) estimates that cotton production was 3.837 million tons in 1960 and 7.748 million tons in 1988.

25. Lipovsky (1995, p. 541) notes that 92% of all Central Asian cotton was sent to the central regions of Russia for processing.

26. As a measure for comparison, Uzbekistan had 42,000 km of main roads and about 90,000 km of local roads, which reached most parts of the country. It had 3500 km of rail lines, but about 1000 km had to be rehabilitated after independence. See pp. 169–170 of IMF 1992.

27. For examples illustrating the difficulty of resolving conflicts among the republics, see Laschenov 1990.

28. In 1988 the name of the all-Union Minvodkhoz was changed to Minvodstroy (Ministry of Water Construction). In 1990 the organization was dissolved.

29. On the role of Minvodkhoz in Uzbekistan see Thurman 1995.

30. Interview with D. Solodennikof, Chief, Department for the Coordination of Water Management Activities and Ecological Cooperation, EC of ICAS, (1995). Also described on. p. 7 of Legal and Institutional Aspects, volume VI, TACIS, Water Resources Management and Agricultural Production in the Central Asians Republics, January 1996.

31. These water distribution calculations rely upon a conglomeration of factors that include, for example, the types of crops grown, norms of water use for each type of crop, soil type, and the net quantity of water that year. Interview with Usman Buranov, Head of Exploitation, Minvodkhoz, Toshkent, Uzbekistan, June 7, 1995.

32. *Irrigatsiia Uzbekistana*, volume 1 (FAN, 1975), pp. 236–267.

33. The chairman of the kolkhoz prepares a plan on water use and distribution for both the vegetation and non-vegetation periods with the hydrotechnician (the Soviet equivalent of the mirab) and the agronomist, which he then brings to the local irrigation administration to negotiate an agreement (dogovor). Interview with Raivodkhoz, Kurgan Tepenkski Rayon, June 14, 1995. The local head also emphasized that farm chairmen always ask for more water than necessary, knowing that they will not receive the total amount requested.

34. Interview with Egal Cohen, Agridev, January 17, 1995, Toshkent, Uzbekistan.

35. Interview with Dr. Konyukov, Vice-Chairman of Goskompriroda, Uzbekistan, February 24, 1995.

36. See e.g. Critchlow 1991a and Fierman 1991.

37. This argument draws on similar observations made by Peter Hauslohner (1987) in the Slavic parts of the Soviet Union.

38. On a similar process in Africa see Bates 1981.

39. A formal definition of rent is "a return received in an activity that is in excess of the minimum needed to attract the resources to that activity" (Milgrom and Roberts 1992, p. 603).

40. Interview with Djalalitdin Sattarov, head of Main Board for External Economic Relations and Personnel for Foreign Activities, Minvodkhoz, Uzbekistan, June 6, 1995.

41. On social control and colonial policies see Migdal 1988, p. 105.

42. For the USSR as a whole, only 19% of the total population worked in agriculture (source: IMF, 1991, cited on p. 34 of Pomfret 1995).

43. Personal observation, Sovkhoz Kelesk, Kazakhstan, June 4, 1992.

44. Aslund (1989, p. 151) infers that the "cotton magnates" in Central Asia had stolen more than 4 billion rubles of which half went into their personal pockets.

45. Rashidov ruled Uzbekistan from 1959 to 1983. During that period he built up a political machine within the republic based upon personal ties and one's regional background.

46. See Craumer 1995. For an attempt to correct the figures on cotton production see pp. 1–2 of Severin 1987.

Chapter 5

1. Micklin (1991, p. 47) points out that many Soviet planners and scientists had anticipated that the expansion of irrigation in the Aral basin in the 1950s and the 1960s would reduce the flow into the sea and had earlier raised their concerns.

2. Quoted in P. Shermukhamedov, Y. Kovalev, and S. Mirzaev, "Komu i zachem eto nuzhno?" *Pravda vostoka,* July 7, 1988.

3. For an extended discussion of attempts to reform the economy see Aslund 1989.

4. For vivid examples of the decline in life expectancy and the rise in infant mortality in the Soviet Union, see Feshbach and Friendly 1992.

5. For example, in 1985 the cotton harvest in Uzbekistan missed its goal by about 300,000 tons despite Moscow's lowering of the target from 6 million to 5.7 million tons (Ann Sheehy, "Uzbek Cotton Harvest Falls Short of Target," Radio Liberty Research Bulletin RL 414/85, December 12, 1985).

6. *Pravda,* January 17, 1988. Some of its main responsibilities included the monitoring and enforcement of environmental regulations, preparation of environmental resolutions, creation of an information system on the state of the environment, development of economic incentives for improving environmental management, and the execution of international agreements with other countries.

7. "Environmental Awakening in the Soviet Union," *Science* 241 (August 1988): 1033–1034.

8. Interview, Tashkent, Uzbekistan, February 24, 1995.

9. Through encouraging debate in public over how to jump start the economy and improve the general standard of living in the country, Gorbachev hoped to foster new ideas that would retard old norms and practices of departmentalism and localism, and, therefore, wipe out political corruption. See "Ecology and the Highest Authority," *International Affairs* (Moscow) (November 1989): 115–129. In regard to Central Asia, the center in Moscow continued to target the cotton sector wherein corruption was rampant.

10. On Chernobyl see Walker 1986.

11. I. Pasevyev, "Semiato bolshaia" *Literaturnaia gazeta,* January 27, 1988; Carley 1989; Annette Bohr, "Infant Mortality in Central Asia," Radio Liberty Research Bulletin RL 352/88, August 4, 1988.

12. Source: author's conversations with women on Sovkhoz Savai, Kurgan Tipenski Region, Uzbekistan, June 1995.

13. Reznichenko 1992 contains a diary of the trip.

14. See e.g. "Voda—eto zhizn Aralu, *Pravda vostoka,* June 25, 1988.

15. Originally published in *Literaturnaia gazeta,* March 18, 1988. See also Fierman 1989, p. 22; Anne Bohr, "Head of Uzbek Writers' Union Criticizes Moscow's Economic Diktat," Radio Liberty Research Bulletin RL 146/88, March 22, 1988.

16. T. Kaipbergenov, "Otvetstvennost pered mirom," *Pravda vostoka,* June 2, 1989.

17. "Molodozh sprashivaet, prezident otvechaet," *Pravda vostoka,* November 29, 1990.

18. FBIS-Sov-89-115, *FBIS Daily Report: Central Eurasia,* June 16, 1989, 49.

19. For local coverage of the riots see *Pravda vostoka* for June 7 and 9, 1989.

20. Many people abide by the officially touted explanation that such conflicts were merely due to hooligans. Interview with Deputy Director and Head of the Division of Water Use at Oblastvodkhoz, Osh, Kyrgyzstan, April 27, 1995. Others consider these conflicts to be much more representative of the overall problem of water sharing in Central Asia. Interview with the Head of the Division of Energy and Natural Resources, International Institute for Strategic Studies (ISS), Bishkek, Kyrgyzstan, February 7, 1995.

21. For a perspective on the roots of the conflict see Tishkov 1995.

22. On opposition to the diversion scheme see Darst 1988.

23. Resolution on the Cessation of Work on the Diversion of the Flow of the Northern and Siberian Rivers, CC CPSU and USSR Council of Ministers, *Pravda,* August 20, 1986.

24. *Pravda vostoka,* June 24, 1990.

25. Previously, President Karimov of Uzbekistan had argued that Central Asia would require additional water supplies to maintain its economy. See *Pravda vostoka* for September 23, 1989.

26. *Pravda,* September 30 1988.

27. In the early 1980s there were some years where no water reached the Aral Sea. See table 1.1.

28. Micklin (1992b, p. 108) points out that this program would not have resurrected the sea but would only have stabilized it at a depth of approximately 34 meters (about 5 meters below the 1989 mark) and an area around 27,000 km2 (13,000 km2 less than in 1989). Moreover, the use of drainage water would have further contributed to the increasing salinity of the sea. In general, most of the polluted water, instead of being directed to the sea, was dumped into land depressions, forming dozens of salt lakes of standing water such as Sarykamysh Lake on the border of Karakalpakstan and Turkmenistan.

29. Micklin believes that the Soviet government was firmly committed to carrying out these plans. See "International and Regional Responses to the Aral Sea Crisis: An Overview of Efforts and Accomplishments," Paper Presented at the Aral Sea Basin Water Management Workshop, Sponsored by the Social Science Research Council, Toshkent, Uzbekistan, May 19–21, 1998.

30. "Ecology and the Highest Authority," *International Affairs* (Moscow) (November 1989): 115–129.

31. After the breakup of the Soviet Union, ISAR changed its name to the Initiative for Social Action and Renewal in Eurasia.

32. Souce: correspondence with William Davoren, Aral Sea International Committee.

33. O. Abdirakhmanov, S. Kamalov, and S. Kabulov, "Otkritoe pisma piyati presidentam," *Sovetskaia Karakalpakiia,* April 25, 1992 (*Svobodniye gori,* June 17, 1992).

34. Before 1960, the average discharge into the sea ranged from 57.5 to 62.5 km^3. Of this, 47–50 km^3 came from the Amu and Syr Daryas, 5–6 from ground water and 5.5–6.5 from precipitation (McKinney and Akmansoy 1998).

35. The Chardara Reservoir in Kazakhstan cannot absorb the increased water runoff from Toktogul. Because of this the water is released into the Arnasai depression across the border in Uzbekistan. See Solodennikof 1996.

36. Interview, ISS, Bishkek, Kyrgyzstan, February 7, 1995.

37. Interview with Parliamentarian Kachikayev, Bishkek, Kyrgyzstan, March 6, 1997.

38. Interviews with Bill Pemberton (Project Manager, European Union, Energy Advisory Group), February 10, 1995, and Orunbek Shamkonov (Energy Specialist, World Bank), March 5, 1997, both at Bishkek, Kyrgyzstan.

39. For a description of the potential development of the hydroelectric sector in Kyrgyzstan, see the final report prepared by Harza Engineering Company for

United States Agency for International Development, dated November 1993 (Harza Engineering Company 1993).

40. Much skepticism has been expressed concerning whether markets exist for the excess power that would be produced. Interview with Rolf Manfred, USAID Energy Consultant, Almaty, Kazakhstan, March 17, 1995.

41. Insofar as Afghanistan is an upstream riparian in the Amu Darya basin, any long-term political solution to the water-sharing situation in Central Asia will have to include it in the negotiation process. The absence of a clear political authority in Afghanistan has precluded its participation in the negotiations, and similar to Tajikistan, decades of political upheaval have prevented it from taking steps to alter the water flows along the Amu Darya.

42. Among the main dams that have attracted international concern is the Sarez Lake Dam ("Tajik Lake Might Cause Disaster," Associated Press, January 19, 1998). See also Salimjon Aioubov, "Central Asia: A Looming Ecological Apocalypse," Radio Free Europe/Radio Liberty, May 23, 1997.

43. This figure is based upon discussions with a lecturer at the Toshkent Institute for Water Engineers and the Mechanization of Agriculture, January 1995.

44. Site visit, 1994. Also see Klötzli 1994; Smith 1995.

45. Although conflicts over water have not erupted since independence, the civil war in Tajikistan crossed over into the Batken region in Kyrgyzstan, resulting in hostage taking and internal conflict in the fall of 1999.

46. "Agreement between the Republic of Kazakhstan, the Republic of Kyrgyzstan, the Republic of Uzbekistan, the Republic of Tajikistan and Turkmenistan on Cooperation in Management, Utilization, and Protection of Water Resources of Interstate Sources." See also p. 8 of volume VI: Legal and Institutional Aspects (Water Resources Management and Agricultural Production in the Central Asian Republics, Toshkent, Uzbekistan, January 1996). This conforms to the recognized principle of international water law in which to use your own in such a manner as not to injure others.

47. The Statute of the Interstate Water Management Coordinating Commission, signed December 5, 1992, regulates the functioning of the IWMCC. For an overview of the legal institutions, see volume VI: Legal and Institutional Aspects (Water Resources Management and Agricultural Production in the Central Asian Republics, Toshkent, Uzbekistan, January 1996).

48. A separate BVO exists for the waters of the Syr Darya below the Chardara Reservoir.

49. This mirrors similar patterns of rapid cooperation among transitional states such as the Visegrad states. Again see Bunce 1997. It is also interesting to note that at this time, the Central Asian leaders were busily signing a series of agreements including border agreements that recognized and reinforced Soviet territorial demarcations. This was part of a general attempt to maintain as much of the status quo in the months after independence.

50. On the role of precedent in environmental negotiations see Barrett 1994, p. 28.

51. Interview with representative of Meredith Jones Group, Toshkent, Uzbekistan, June 2, 1992.

52. Source: conversations with several representatives from the Eisenberg project from Israel, Toshkent, Uzbekistan, June 1992. Eisenberg had projects in Uzbekistan, Tajikistan, and Shymkent region in Kazakhstan.

53. Interview at Ministry of Water Resources (Minvodkhoz), Toshkent, Uzbekistan, May 13, 1993.

54. Interview with Director and Deputy Director, International Institute for Strategic Studies, Kyrgyzstan, January 31, 1995.

55. Interview with Sayfulla Persheyev, representative of Uzbekistan's State Oil and Gas Company, Uzbekneftegas, Toshkent, Uzbekistan, March 19, 1997.

56. In addition, in place of Minvodkhoz, Kazakhstan adopted a State Committee on Water Resources. Moreover, with time, many of the countries began to merge their water ministries with their agricultural ministries.

57. Interview with Director, International Institute for Strategic Studies, Kyrgyzstan, January 31, 1995.

58. Volume VI: Legal and Institutional Aspects.

59. Ibid., p. 11.

60. *Kazakhstanskaia pravda,* December 24, 1992.

61. It was impossible for me to visit the Aral Sea in the summer of 1992, owing to many government obstacles, but in 1994 I was able to travel to the Aral Sea region unhindered.

62. Interview with representative from Birlik, June 17, 1992, Toshkent Uzbekistan.

Chapter 6

1. *Hokim* is the Uzbek term for regional leader. In Kazakh, the equivalent is *akim.*

2. There was a growing fear in Western policy circles that Islamic fundamentalism might spread from Iran.

3. With US assistance through the Cooperative Threat Reduction program, Kazakhstan agreed early on to dismantle its nuclear weapons and remove enriched uranium from its territory (Jones and McDonough 1998, pp. 79–88).

4. The region's proven oil reserves are estimated at 18–35 billion barrels and proven natural gas reserves are even more significant, estimated at 236–337 trillion cubic feet. US Department of Energy, "Caspian Sea Region," June 2000 (http://www.eia.doe.gov/emeu/cabs/caspian.html).

5. Interview with Werner Roeder, Toshkent, Uzbekistan, March 20, 1997.

6. Kaipbergenov, "Tri uroka Arala."

7. "International and Regional Responses to the Aral Crisis: An Overview of Efforts and Accomplishments," Paper Presented at the Aral Sea Basin Water Management Workshop Sponsored by the Social Science Research Council, Toshkent, Uzbekistan, May 19–21, 1998, p. 2.

8. Only with the collapse of the Soviet Union were scientists and policy makers able to expose the severity of the nuclear legacy. See Bradley 1997. The Central Asian governments are still reluctant to fully reveal the damage caused by mining for uranium. Even in Russia, nuclear related issues are highly sensitive and remain classified. For example, Aleksandr Nikitin was charged for espionage for providing information on nuclear waste and submarines in northern Russia to the Norwegian NGO Bellona.

9. In *Earth in the Balance* Gore makes numerous references to his trip to the Aral Sea.

10. Unclassified memorandum on US Aral Sea strategy from Strobe Talbott, dated July 22, 1993, cited in Environmental Policy and Technology (EPT) Project Summary: US Aral Sea Program Overview, Prepared for Regional Mission for Central Asia, US Agency for International Development, October 2, 1996.

11. Interview, Almaty, Kazakhstan, March 17, 1995.

12. The five Central Asian states were admitted as members into the United Nations on March 2, 1992.

13. Until the entrance of Azerbaijan, Turkmenistan, Uzbekistan, Tajikistan, and Kyrgyzstan as members on February 17, 1992, this organization, consisting of Pakistan, Turkey, Iran, and Afghanistan, was largely defunct.

14. Interview with Steven Wolfe, ISAR representative in Central Asia, Almaty, Kazakhstan, March 22, 1995.

15. Personal correspondence, June 1998.

16. For a summary of the events leading up to the advent of the World Bank, Aral Sea Program, see Aral Sea Program—Phase 1, Briefing Paper for the Proposed Donors Meeting to be Held on June 23–24, 1994 in Paris, World Bank/UNDP/UNEP, May 1994.

17. Personal communication with World Bank water economist, November 1993.

18. Some consultants in informal discussions have noted that the World Bank should have refrained from pushing for a new institutional structure, which in the end "was only cosmetic."

19. Four of the principles of international water law are to inform and consult with water-sharing neighbors before taking actions that may affect them, to regularly exchange hydrological data, to avoid causing substantial harm to other water users, and to allocate water from a shared river basin reasonably and equitably. For a discussion of international water law see McCaffrey 1993.

20. *Kazakhstanskaia pravda,* March 30, 1998.

21. This agreement and other related documents provide the basis for ICAS, EC-ICAS, and the ICSDSTEC. The statutes laid out the specific role of these new

organizations and the relationship among them. For an overview of all the agreements see volume VI: Legal and Institutional Aspects.

22. Volume VI: Legal and Institutional Aspects, p. 1. One of the TACIS legal advisors found that these institutions "are regulated by separate statutes that are not completely streamlined as to the institutions' respective functions." Hence, it was difficult to distinguish between regulatory and development institutions. For further clarification see Caponera 1995 and Nanni 1996.

23. Volume VI: Legal and Institutional Aspects, p. 9.

24. V. Kirchev, "Seminar in Washington," *Narodnoe slovo,* May 27, 1993.

25. On p. 1 of World Bank 1995b it is noted that two of the projects were later merged, leaving 18 individual projects.

26. *Panorama* (newspaper in Kazakhstan), January 15, 1994.

27. Aral Sea Program—Phase 1, Briefing Paper for the Proposed Donors Meeting, May 1994, p. ii.

28. This estimate was for the first 3 years (World Bank 1995b, p. iii).

29. This project received early GEF funding ($375,000) and was administered by the World Bank in close coordination with the European Union's WARMAP project.

30. Aral Sea Program—Phase 1, Briefing Paper for the Proposed Donors Meeting, p. vi.

31. Interview, World Bank, Toshkent, Uzbekistan, January 18, 1995.

32. Similarly, on the changing role of donor institutions in Eastern Europe and how they set the agenda, see Connolly, Gutner, and Bedarff 1996 and Gutner 1999.

33. Kapuscinski (1994, pp. 254–264) brilliantly captures the Soviet mentality of seeking technical fixes or solutions from the outside rather than ever having to consider hard-budget constraints or environmental and social effects of policy choices. Even after the Soviet Union collapsed, research institutes continued to work on the Siberian river diversion project. Also see Iskandarov 1994.

34. Meeting with the Institute of Water Problems as part of a National Academy of Sciences research trip, August 17, 1994.

35. Interview with Raffik S. Saifulin, Deputy Director and Nodir S. Sultanov, Professor, Toshkent Uzbekistan, January 17, 1995.

36. P. Shermukhamedov, "Vodi Kaspiya—Aralu?" *Narodnoe slovo,* April 25, 1995.

37. At that point, the World Bank, UNDP, and EU-TACIS signed a memorandum of understanding to avoid duplication of projects and to cooperate with one another to reach the overall objectives of the Aral Sea Basin Program.

38. Executive Summary, TACIS, Water Resources Management and Agricultural Production in the Central Asians Republics, Toshkent, January 1996.

39. Volume VI: Legal and Institutional Aspects, p. 2. TACIS advisors are to assist with the "drafting of international (interstate) agreements on policies and strategies on water and land resources, their use, management, protection and apportionment; the drafting of national and intergovernmental legal and normative acts based on interstate agreements."

40. For an overall assessment see Boisson de Chazournes 1998.

41. Dates for the adoption of new water codes and laws are the following: Kazakhstan, March 11, 1993; Kyrgyzstan, January 14, 1994; Tajikistan, December 27, 1993; Uzbekistan, March 6, 1993.

42. This was one of the recommendations put forth in Caponera 1995.

43. The World Bank, the UNDP, and UNEP signed a Memorandum of Understanding on collaboration among them to define the role of the three organizations in respect to the Aral Sea Program on November 30, 1994.

44. President Niyazov of Turkmenistan did not attend the meeting, but he later signed the declaration.

45. A copy of the Nukus Declaration is included in World Bank 1996b.

46. Interview, Toshkent, Uzbekistan, February 26, 1997.

47. Also see the other case studies in Greenberg, Barton, and McGuinness 2000.

48. Source: email from Peter Whitford of Aral Sea Basin Program, World Bank, Washington, October 19, 1997.

49. Interview, IFAS, Almaty, Kazakhstan, March 29, 1995.

50. Interview, Institute of Geography, Almaty, Kazakhstan, March 21, 1995.

51. During the Soviet period, the Institute for Irrigation Engineers trained students from all over Central Asia and from Africa, Asia, and Latin America.

52. Interview, Toshkent, Uzbekistan, February 24, 1995.

53. Interview, IFAS, Almaty, Kazakhstan, March 29, 1995.

54. Personal observation from informal conversations with local scientists at both these institutes in 1994–95.

55. Protocol of the Workshop of the Program Group 1, February 7–9, 1995, Toshkent.

56. Members of the Leading Group for Implementation Works of Phase 1 Program 1, Approved by the Chairman of the EC-ICAS, A. Ilamanov, December 12, 1994.

57. Members of the Leading Group for Implementation Works of Phase 1 Program 3, Approved by the Chairman of the EC-ICAS, A. Ilamanov, December 12, 1994.

58. Interview with Arrigo Di Carlo, February 26, 1997.

59. Interview with Leonid Dmitriev, Chief Engineer, Kazgiprovodkhoz, Almaty, Kazakhstan, March 7, 1995.

60. Interview with Paul Dreyer, March 17, 1995. The Toshkent workshop had 73 attendees, and 19 papers were presented.

61. Environmental Policy and Technology (EPT) Project Summary: US Aral Sea Program Overview, Prepared for Regional Mission for Central Asia, US Agency for International Development, October 2, 1996.

62. For an overview see World Bank 1994a. Tajikistan is particularly worried about a potential rupture of Lake Sarez, an artificial dam created by an earthquake and landslides in 1911.

63. Interview with representatives from Eisenberg, Toshkent, Uzbekistan, June 1992. Eisenberg had established cotton projects on Sovkhoz Kelesk in Shymkent and on Sovkhov Savai in the Fergana Valley.

64. Interview with World Bank agricultural economist, Toshkent, Uzbekistan, May 4, 1995. After the project failed to produce the desired results, the Ministry of Internal Affairs initiated an investigation to establish how it was possible for the hokim in Andijon to negotiate such a deal outside the Ministry of Agriculture. This is just one example of the way in which the hokims could influence politics outside of the normal ministerial channels and also impede international efforts to introduce economic reform.

65. Outside the agricultural sector, one of the starkest images of the influence of foreign actors is the introduction of new laws on Foreign Investment, Licensing, Banking and Environmental Regulations in Kazakhstan.

66. The resistance to dismantling cotton monoculture is discussed in detail in chapter 7.

67. Interview with Michael Scanlan, US Embassy, Bishkek, Kyrgyzstan, February 9, 1997. He suggested that when the donors leave, this will most likely "pull the rug out" from under Kyrgyzstan. See also Olcott 1996, pp. 87–112.

68. Reform in the agricultural sector began in 1992. Yet, by early 1993 only 165 state or collective farms (out of a total of 480) had been reorganized or privatized. At that time, agricultural reform was suspended but then later reinstated in 1994 when land reform was put in the hands of the Ministry of Agriculture. Source: author's conversation with John Scherring, TACIS, Bishkek, Kyrgyzstan, February 8, 1995. See also Paul Munro-Faure, *Interim Report on Land Reform Issues, May-June, 1994,* ULG Consultants Ltd, Ministry of Agriculture of the Government of Kyrgyzstan and EC-TACIS Sectoral Support Program.

69. Projects were being carried out in Osh and in Naryn, Kyrgyzstan. Interview with Pierre Mantion, TACIS-Livestock Project, Osh, Kyrgyzstan, June 27, 1995.

70. On agricultural reform see Spoor 1995.

71. EPT Memorandum on Water Management Study Tour, May 27, 1994.

72. Interview with Kadir Ergashev, UNDP, Toshkent, Uzbekistan, January 13, 1995.

73. For an overview of the way that the Uzbekistan government has suppressed any form of dissent, see Polat 1995.

74. Only President Askar Akaev was not a former republican party secretary.

75. United Nations Development Programme Regional Project, Aral Sea Basin Capacity Development, Aral Sea Basin NGO Directory (Aral Sea Basin Capacity

Development Project, Toshkent, 1996). It was noted on p. 9 of World Bank 1996b that UNICEF was conducting programs of similar size in the disaster zones of Kazakhstan and Turkmenistan.

76. Personal communication, Daene McKinney, USAID consultant, May 23, 1999.

77. In order to demonstrate a commitment to the newly independent Central Asian states, the US government devised several small but visible projects. In October 1993, Secretary of State Warren Christopher announced a $15 million program to help alleviate environmental conditions in Aral Sea region. In June 1994, an additional $7 million was pledged for these projects.

78. Environmental Policy and Technology (EPT) Project Summary, October 2, 1996.

79. Interview with Paul Dreyer, March 17, 1995.

80. Interview, UNDP, Toshkent, Uzbekistan, February 26, 1997. They have only spent $120,000 on the hand pumps, but this has been more successful than their bigger programs.

81. I observed this on visits to several state and collective farms in Kazakhstan and Uzbekistan where such projects are underway.

82. This sentiment was articulated in a formal response to Philip Micklin's paper at the SSRC Water Conference by Oleg Tsaruk, Law and Environment Eurasia Partnership (LEEP)/Aral Sea International Committee, May 19–21, 1998, Toshkent, Uzbekistan.

83. *Ecostan News* and ISAR's *Surviving Together* followed the development of such environmental NGOs in Central Asia. *Surviving Together* was later replaced by *Give and Take.*

84. Cited in *Ecostan News*, February 1997 and June 1997.

85. Personal correspondence with Bill Davoren, June 1998. In contrast, some representatives from the World Bank have informally expressed their reservations in regard to having NGOs participate in the meetings and seminars.

86. The representative from MSF managed to attend the meeting on his own initiative.

87. ISAR, Summary Report to the Turner Foundation: Supporting Alternative Energy in Central Asia, February 1999.

88. Formal response to Philip Micklin's paper at the SSRC Water Conference, May 19–21, 1998, Toshkent, Uzbekistan.

89. See also Prosser 2000.

Chapter 7

1. Weinthal 2001 is an extended version of this chapter.

2. Although both Turkmenistan and Uzbekistan possessed substantial reserves of oil and gas reserves that could serve as an alternative export commodity, it

could not bring the immediate financial benefits of cotton. Thus, they chose to delay the development and exploitation of these energy resources. See Jones Luong and Weinthal 2001.

3. Interview, Toshkent, March 21, 1997.

4. The legal report was by Dante Caponera (1995).

5. The Program Group worked closely with the Scientific Information Center of the ICWC. The national working groups were led by Kirgyzgiprovodkhoz, TajNIIGiM, Turkmengiprovodkhoz, Uzvodprojekt, and Kazkgiprovodkhoz. The coordinators were Viktor Dukhovny, Janusz Kindler (WB), and Arrigo Di Carlo (WARMAP).

6. Interview with TACIS consultant, June 22, 1999.

7. More so, the heads of state essentially adopted all of Caponera's main recommendations that had initially upset the Central Asian water establishment. As one World Bank consultant said, "time heals all wounds."

8. It was agreed that the leadership would be rotated every 2 years. In 1999, IFAS was transferred to Ashgabat.

9. Resolution of the International Fund for the Aral Sea on the Establishment of the Executive Committee of the International Fund for the Aral Sea, March 20, 1997.

10. Interview, Almaty, March 13, 1997.

11. Interview with TACIS consultant, June 22, 1999.

12. Ibid.

13. Interview withWerner Roeder, Toshkent, March 20, 1997.

14. Interview, Almaty, March 13, 1997.

15. Letter from Jogorku Kenesh to Michael Rathnam, World Bank, Bishkek, Kyrgyzstan, 28.06.97, No. 01–12/94.

16. Protocol of the Meeting of Representatives of Fuel-Energy and Water Management Complexes of Kazakhstan, Kyrgyzstan, and Uzbekistan on Problem of the Toktogul Cascade Water-Energy Resources Use in 1996, ICWC Bulletin 11, November 1996; Agreement between the Governments of the Kyrgyz Republic and the Republic of Uzbekistan on the Question of the Use of Hydro-electric Resources of the Naryn-Syr Darya Hydro-Electric Power Station, 1996.

17. Sander Thoenes, "Central Asians Reach Common Ground over Water," *Financial Times,* April 9, 1996.

18. As water is used in the winter months for electricity production even though this is a non-consumptive use, the water never reached the Aral Sea. Instead, it was diverted to depressions downstream to avoid flooding.

19. In addition, USAID had other reasons for lending additional support to Kyrgyzstan and Kazakhstan over Uzbekistan. It considered the ICKKU not to be an Uzbekistan-dominated organization. It had closer relations with both Kyrgyzstan and Kazakhstan than with Uzbekistan since they had made greater progress

toward opening up their societies to markets and democracy. USAID had set up its hub in Almaty to assist the transition. Thus, whereas the World Bank's Aral Sea Basin Program and the European Union's TACIS headquarters were situated in Toshkent, USAID could offer a counter balance. Moreover, USAID was keen on promoting privatization and pricing policies in both the water and energy sectors, which Uzbekistan was resisting.

20. Source: conversation with Barbara Britton, USAID-EPT Project, Almaty, Kazakhstan, March 10, 1997.

21. Interview, Almaty, Kazakhstan, March 11, 1997.

22. Ibid.

23. Cable from US Embassy in Almaty Kazakhstan, November 27, 1996 (AID).

24. Cable from US Embassy in Almaty Kazakhstan, January 6, 1997 (AID).

25. Minutes of the Energy and Water Roundtable Third Session, The Efficient Use of the Naryn-Syr Darya Cascade Water Storage Agreement in Principle, Ysyk-Kol, Kyrgyzstan, July 1–4, 1997.

26. Cable from US Embassy in Almaty Kazakhstan, September 22, 1997 (AID).

27. Under the recommended draft agreement, for the next 5 years Kyrgyzstan agreed to provide summertime releases of 3.25 km^3 of water and to supply 1.1 billion kWh of electricity to Kazakhstan. Kazakhstan would either pay for the power provided at the rates current for the signing of the agreement, or in exchange would supply Kyrgyzstan 1.1 billion kWh of electricity in winter or would provide delivery of other fuel resources in equivalent volumes depending on the agreement with Kyrgyzstan. Uzbekistan and Kyrgyzstan—Kyrgyzstan agreed to provide summertime releases of 3.25 km^3 of water from Toktogul due to the water regime of Uzbekistan in the vegetation period and to supply 1.1 billion kWh of electricity to Uzbekistan. Uzbekistan would deliver 500 million m3 of natural gas and 400 million kWh of electricity (at a price of $0.4/kWh) in autumn and winter. The total estimated value of the exchange was $48.5 million. Source: minutes of Energy and Water Roundtable, third session ("The Efficient Use of the Naryn–Syr Darya Cascade Water Storage Agreement in Principle"), Ysyk-Kol, Kyrgyzstan, July 1–4, 1997.

28. When Tajikistan joined the ICKKU, its name was changed to ICKKTU.

29. USAID's water program was under CH2M-Hill International Services and the energy program under Hagler Bailly.

30. USAID Memorandum, Hagler Bailly, Bishkek, Kyrgyzstan, September 28, 1997.

31. Interview, Almaty, Kazakhstan, March 12, 1997.

32. Interview, June 22, 1999.

33. Interview, Almaty, March 13, 1997.

34. Only in 1996 did the agriculture ministry and water ministry merge into one ministry in Uzbekistan. This followed a trend that had been taking place in Kyrgyzstan and Kazakhstan.

35. During this period, it exported 79–95% of its cotton. For Uzbekistan's cotton data, see http://www.fas.usda.gov.

36. Human Development under Transition—Uzbekistan (http://www.undp.org:80/rbec/nhdr/1996/summary/uzbekistan.htm).

37. Land Reform and Farm Restructuring Policy Guidelines, Food and Agricultural Policy Advisory Unit, Tacis Programme of the European Commission, Toshkent Uzbekistan, December 1998, p. 1.

38. In Uzbekistan, the Ministry of Foreign Economic Relations limited access to foreign cotton buyers since cotton sales after independence required its signature and the president's.

39. Ron Synovitz, "Uzbekistan: Little Progress Seen in Agricultural Reforms," Radio Free Europe/Radio Liberty, February 25, 1997.

40. Land Reform and Farm Restructuring Policy Guidelines, December 1998, p. 16.

41. Ibid. If children were eligible for a land share, as in Kyrgyzstan, then the land share would be 0.16 hectare.

42. In the high-population-density areas, such as the Fergana Valley, the plots are smaller than in the low-population-density areas, such as in Karakalpakstan

43. Synovitz, "Uzbekistan." I observed the same situation on state farms in Turkmenistan in August 1994.

Chapter 8

1. Consensus does not exist on how to measure effectiveness. Young (1994a) has pointed out that effectiveness can be measured in at least six different ways. Other studies of effectiveness include Bernauer 1995; Haas, Keohane, and Levy 1993; and Victor, Raustiala, and Skolnikoff 1998.

2. See also IBRD 1997. In this report, the World Bank acknowledges the role of the state, pulling back from its more extreme neoliberalist perspective.

3. In a study of the impact of IOs on the nation-state, McNeely (1995, p. 58) finds that "a fundamental aspect of sovereignty is the absolute competence of a state to perform acts and make treaties and agreements in the international arena."

4. Interview with Konyukhov, February 24, 1995; interview with Turemuratov, March 29, 1995.

5. The Convention on the Law of the Non-navigational Uses of International Watercourses, 1997 was adopted by the UN General Assembly in resolution 51/229 of May 21, 1997.

6. On institutional isomorphism see DiMaggio and Powell 1991.

7. For example, Uzbekistan (June 20, 1993), Kazakhstan (May 17, 1995), Turkmenistan (June 5, 1995), and Tajikistan (July 1, 1998) signed the Framework

Convention on Climate Change. Similarly, all five Central Asian states have joined the Convention on Biological Diversity.

8. Interview with UNDP, February 26, 1997.

9. Interview with Michael Boyd, Harvard Institute for International Development, Almaty, Kazakhstan, March 27, 1995.

10. Akhmal Karimov, History of Irrigation in Uzbekistan and Present Problems (unpublished paper, 1997).

11. Correspondence, Cassandra Cavanaugh, Human Rights Watch.

12. Interview with Vadim Igorevich Antonov, Vodnoproekt, Toshkent, Uzbekistan, January 24, 1995.

13. Interview, Osh Oblastvodkhoz, April 27, 1995.

14. Ibid. An ethnic Uzbek member of the Kyrgyzistani parliament also reiterated this opinion.

15. Interview, EU-TACIS, Bishkek, Kyrgyzstan, March 5, 1997.

16. For details of the Caspian Environment Program, see Evans and Kinman 2001, p. 39.

17. Again, on state formation in Western Europe see Tilly 1975.

18. On the difficulty of undertaking economic reform during a political transition, see Przeworski 1991.

19. On the preference for stability see Starr 1996.

20. Moreover, some members of the international community actually endorsed the annulment of the 1994 parliamentary elections as a "democratic" move. In March 1995, the Constitutional Court of Kazakhstan declared these previous elections to be unconstitutional citing that they did not conform to the principle of one person-one vote guaranteed in Kazakhstan's constitution. President Nazarbayev, in turn, used this ruling as an opportunity to dissolve the increasingly hostile and independent parliament that was elected in March 1994 and to rule by decree in the interim.

21. On oil rents and taxation see Karl 1997.

22. Interview with donor, Ramallah, Palestinian Authority, April 19, 2000.

References

Abramov, I. 1916. "Polozhenie ob upravlenie Turkestanskogo Kraia." Toshkent.

Allworth, Edward, ed. 1994. *Central Asia: 130 Years of Russian Dominance, A Historical Overview.* Third edition. Duke University Press.

Aminova, R. 1974. *Changes in Uzbekistan's Agriculture (1917–1929).* Nauka.

Anderson, Mary. 1999. *Do No Harm: How Aid Can Support Peace—Or War.* Lynne Rienner.

Andrianov, B., et al. 1991. *Aralski krizis.* Institute of Ethnography and Anthropology of the Soviet Union.

Antonov, Vadim. 1995. "Kakou dolzhna byt napravlennost razrabotki programmy I." Prepared for World Bank's Program I. Toshkent, January.

Aslund, Anders. 1989. *Gorbachev's Struggle for Economic Reform.* Pinter.

Atkin, Muriel. 1993. "Tajikistan: Ancient Heritage, New Politics." In *Nations and Politics in the Soviet Successor States,* ed. I. Bremmer and R. Taras. Cambridge University Press.

Axelrod, Robert. 1984. *The Evolution of Cooperation.* Basic Books.

Bacon, Elizabeth. 1966. *Central Asians under Russian Rule: A Study in Culture Change.* Cornell University Press.

Baechler, Günther, and Kurt Spillman, eds. 1996. *Environmental Degradation as a Cause of War,* volumes 2 and 3. Rüegger.

Baldwin, David, ed. 1993. *Neorealism and Neoliberalism: The Contemporary Debate.* Columbia University Press.

Barkin, Samuel, and George Shambaugh, eds. 1999. *Anarchy and the Environment: The International Relations of Common Pool Resources.* State University of New York Press.

Barnett, Michael, and Martha Finnemore. 1999. "The Politics, Power, and Pathologies of International Organizations." International Organization 53: 699–732.

Barrett, Scott. 1990. "The Problem of Global Environmental Protection." *Oxford Review of Economic Policy* 6: 68–79.

Barrett, Scott. 1994. Conflict and Cooperation in Managing International Water Resources. Policy Research Working Paper 1303, World Bank.

Bartold, V. 1914. *K istorii orosheniia Turkestana.* Sankt Petersburg.

Bartold, V. 1927. *Istoriia kulturnoi zhizni Turkestana.* Academy of Sciences of the USSR.

Bates, Robert. 1981. *Markets and States in Tropical Africa.* University of California Press.

Beaumont, Peter, Michael Bonine, and Keith McLachlan, eds. 1989. *Qanat, Kariz and Khattara: Traditional Water Systems in the Middle East and North Africa.* Middle East and North African Studies Press.

Bedford, Dan. 1996. "International Water Management in the Aral Sea Basin." *Water International* 21: 63–69.

Benedick, Richard. 1991. *Ozone Diplomacy.* Harvard University Press.

Bernauer, Thomas. 1995. "The Effect of International Environmental Institutions: How We Might Learn More." *International Organization* 49: 351–377.

Bernauer, Thomas. 1997. "Managing International Rivers." In *Global Governance,* ed. O. Young. MIT Press.

Birnie, Patricia. 1992. "International Environmental Law: Its Adequacy for Present and Future Needs." In *The International Politics of the Environment,* ed. A. Hurrell and B. Kingsbury. Clarendon.

Birnie, Patricia, and Alan Boyle. 1993. *International Law and the Environment.* Clarendon.

Bohr, Annette. 1989. "Health Catastrophe in Karakalpakstan." *Report on the USSR,* July 21: 37–38.

Boisson de Chazournes, Laurence. 1998. "Elements of a Legal Strategy for Managing International Watercourses: The Aral Sea Basin." In International Watercourses: Enhancing Cooperation and Managing Conflict (World Bank Technical Paper 414).

Boycko, Maxim, Andrei Shleifer, and Robert Vishny. 1997. *Privatizing Russia.* MIT Press.

Bradley, Don. 1997. *Behind the Nuclear Curtain: Radioactive Waste Management in the Former Soviet Union.* Battelle Press.

Buck, Susan, Gregory Gleason, and Mitchell Jofuku. 1993. "'The Institutional Imperative': Resolving Transboundary Water Conflict in Arid Agricultural Regions of the United States and the Commonwealth of Independent States." *Natural Resources Journal* 33: 595–628.

Bunce, Valerie. 1997. "The Visegrad Group: Regional Cooperation and European Integration in Post-Communist Europe." In *Mitteleuropa: between Europe and Germany,* ed. P. Katzenstein. Berghahn Books.

Bunce, Valerie. 1999. *Subversive Institutions: The Design and the Destruction of Socialism and the State.* Cambridge University Press.

Caponera, Dante. 1985. "Patterns of Cooperation in International Water Law: Principles and Institutions." *Natural Resources Journal* 25: 563–587.

Caponera, Dante. 1992. *Principles of Water Law and Administration: National and International.* A. A. Balkema.

Caponera, Dante. 1995. Legal and Institutional Framework for the Management of the Aral Sea Basin Water Resources. Report prepared for EU-TACIS Program on Water Resources Management and Agricultural Production in the Central Asian Republics, Toshkent, April.

Carley, Patricia. 1989. "The Price of the Plan." *Central Asian Survey* 8: 1–38.

Carlisle, Donald. 1991. "Uzbekistan and the Uzbeks." *Problems of Communism,* September-October: 23–44.

Carrère d'Encausse, Hélène. 1994. "Systematic Conquest, 1865 to 1884." In *Central Asia: 130 Years of Russian Dominance,* ed. E. Allworth. Duke University Press.

Chalidze, Francheska. 1992. "Aral Sea Crisis: A Legacy of Soviet Rule." *Central Asia Monitor* 1: 30–36.

Chambers, Robert. 1980. "Basic Concepts in the Organization of Irrigation." In *Irrigation and Agricultural Development in Asia: Perspectives from the Social Sciences,* ed. E. Coward Jr. Cornell University Press.

Chayes, Abram, and Antonia Handler Chayes, eds. 1996. *Preventing Conflict in the Post-Communist World: Mobilizing International and Regional Organizations.* Brookings Institution.

Clague, Christopher, and Gordon Rausser. 1992. *The Emergence of Market Economies in Eastern Europe.* Blackwell.

Clark, Ann Marie, Elisabeth Friedman, and Kathryn Hochstetler. 1998. "The Sovereign Limits of Global Civil Society: A Comparison of NGO Participation in UN World Conferences on the Environment, Human Rights, and Women." *World Politics* 51 (October): 1–35.

Clarke, Robin. 1993. *Water: The International Crisis.* MIT Press.

Colton, Timothy, and Robert Legvold, eds. 1992. *After the Soviet Union: From Empire to Nations.* Norton.

Conca, Ken. 1994. "Rethinking the Ecology-Sovereignty Debate." *Millennium* 23: 701–711.

Connolly, Barbara, Tamar Gutner, and Hildegard Bedarff. 1996. "Organizational Inertia and Environmental Assistance to Eastern Europe." In *Institutions for Environmental Aid,* ed. R. Keohane and M. Levy. MIT Press.

Cornes, Richard, and Todd Sandler. 1986. *The Theory of Externalities, Public Goods, and Club Goods.* Cambridge University Press.

Coward, E. Walter, Jr. 1980. "Irrigation Development: Institutional and Organizational Issues." In *Irrigation and Agricultural Development in Asia: Perspectives from the Social Sciences,* ed. E. Coward Jr. Cornell University Press.

Craumer, Peter. 1995. *Rural and Agricultural Development in Uzbekistan.* Royal Institute of International Affairs.

Critchlow, James. 1991a. *Nationalism in Uzbekistan.* Westview.

Critchlow, James. 1991b. "Prelude to 'Independence': How the Uzbek Party Apparatus Broke Moscow's Grip on Elite Recruitment." In *Soviet Central Asia: The Failed Transformation,* ed. W. Fierman. Westview.

Critchlow, James. 1994. "Nationalism and Islamic Resurgence in Uzbekistan." In *Central Asia,* ed. H. Malik. St. Martin's Press.

Critchlow, James. 1995. "Central Asia: How to Pick Up the Pieces?" In *Environmental Security and Quality After Communism: East Europe and the Soviet Successor States,* ed. J. DeBardeleben and J. Hannigan. Westview.

Crow, Ben, Alan Lindquist and David Wilson. 1995. *Sharing the Ganges: The Politics and Technology of River Development.* Sage.

Dabelko, Geoffrey, and David Dabelko. 1995. "Environmental Security: Issues of Conflict and Redefinition." *Environmental Change and Security Project Report* 1, spring: 3–13.

Daly, Herman. 1996. *Beyond Growth.* Beacon.

Darst, Robert. 1988. "Environmentalism in the USSR: The Opposition to the River Diversion Projects." *Soviet Economy* 4: 223–252.

Darst, Robert. 1997. "The Internationalization of Environmental Protection in the USSR and its Successor States." In *The Internationalization of Environmental Protection,* ed. M. Schreurs and E. Economy. Cambridge University Press.

Darst, Robert. 2001. *Smokestack Diplomacy: Cooperation and Conflict in East-West Environmental Politics.* MIT Press.

Davis, Peter. 1971. "The Law's Response to Conflicting Demands for Water: the United States and the Soviet Union." In *Water Resources Law and Policy in the Soviet Union,* ed. I. Fox. University of Wisconsin Press.

Dawson, Jane. 1996. *Eco-Nationalism.* Duke University Press.

DeSombre, Elizabeth, and Joanne Kauffman. 1996. "The Montreal Protocol Multilateral Fund: Partial Success Story." In *Institutions for Environmental Aid,* ed. R. Keohane and M. Levy. MIT Press.

Deudney, Daniel. 1990. "The Case Against Linking Environmental Degradation and National Security." *Millennium* 19, winter: 461–476.

DiMaggio, Paul, and Walter Powell. 1991. "The Iron Cage Revisited: Institutional Isomorphism and Collective Rationality in Organizational Fields." In *The New Institutionalism in Organizational Analysis,* ed. W. Powell and P. DiMaggio. University of Chicago Press.

Dukhovny, V., and R. Razakov. 1988. "Aral: Gliadia Pravde v Glaza." *Melioratsiia i vodnoe khoziaistvo* 9: 27–32.

Ebel, Robert, and Rajan Menon, eds. 2000. *Energy and Conflict in Central Asia and the Caucasus.* Rowman & Littlefield.

Elhance, Arun P. 1999. *Hydropolitics in the Third World: Conflict and Cooperation in International River Basins.* United States Institute of Peace.

Ellis, William. 1990. "A Soviet Sea Lies Dying." *National Geographic* (February): 73–92.

Elpiner, Leonid. 1999. "Public Health in the Aral Sea Coastal Region and the Dynamics of Changes in the Ecological Situation." In *Creeping Environmental Problems and Sustainable Development in the Aral Sea Basin,* ed. M. Glantz. Cambridge University Press.

Esty, Daniel, Jack Goldstone, Ted Robert Gurr, Pamela Surko, and Alan N. Unger. 1995. State Failure Task Force Report. Prepared for CIA Directorate of Intelligence, November 30.

Evans, Amy, and Michelle Kinman. 2001. "Caspian Environment Programme: Thematic Areas to Define Activities." *Give and Take* 3, winter: 39.

Evans, Peter. 1993. "Building an Integrative Approach to International and Domestic Politics." In *Double-Edged Diplomacy: International Bargaining and Domestic Politics,* ed. P. Evans et al. University of California Press.

Evans, Peter, Harold Jacobson, and Robert Putnam, eds. 1993. *Double-Edged Diplomacy: International Bargaining and Domestic Politics.* University of California Press.

Fairman, David. 1996. "The Global Environment Facility: Haunted by the Shadow of the Future." In *Institutions for Environmental Aid,* ed. R. Keohane and M. Levy. MIT Press.

Falkenmark, Malin. 1986. "Fresh Waters as a Factor in Strategic Policy and Action." In *Global Resources and International Conflict,* ed. A. Westing. Oxford University Press.

Feeny, David, Fikret Berkes, Bonnie McCay, and James M. Acheson. 1990. "The Tragedy of the Commons: Twenty-Two Years Later." *Human Ecology* 18: 1–19.

Ferguson, James. 1994. *The Anti-Politics Machine: "Development." Depoliticization, and Bureaucratic Power in Lesotho.* University of Minnesota Press.

Feshbach, Murray, and Alfred Friendly Jr. 1992. *Ecocide in the USSR.* Basic Books.

Fierman, William. 1989. "*Glasnost'* in Practice: The Uzbek Experience." *Central Asian Survey* 8: 1–45.

Fierman, William, ed. 1991. *Soviet Central Asia: The Failed Transformation.* Westview.

Fitzmaurice, John. 1996. *Damming the Danube: Gabcikovo and Post-Communist Politics in Europe.* Westview.

Fox, Jonathan, and L. David Brown. 1998. *The Struggle for Accountability: The World Bank, NGOs, and Grassroots Movements.* MIT Press.

Frederick, Kenneth. 1996. "Water as a Source of International Conflict." *Resources for the Future,* spring: 9–12.

Frey, Frederick. 1993. "The Political Context of Conflict and Cooperation over International River Basins." *Water International* 18: 54–68.

Fukuyama, Francis. 1989. "The End of History?" *The National Interest* 16, summer: 3–18.

Giddens, Anthony. 1987. *The Nation-State and Violence*. University of California Press.

Glantz, Michael, ed. 1999. *Creeping Environmental Problems and Sustainable Development in The Aral Sea Basin.* Cambridge University Press.

Glantz, Michael, Alvin Rubinstein, and Igor Zonn. 1994. "Tragedy in the Aral Sea Basin: Looking Back to Plan Ahead?" In *Central Asia: Its Strategic Importance and Future Prospects,* ed. H. Malik. St. Martin's Press.

Gleason, Gregory. 1990a. "Marketization and Migration: The Politics of Cotton in Central Asia." *Journal of Soviet Nationalities* 1, summer: 66–98.

Gleason, Gregory. 1990b. "Nationalism or Organized Crime? The Case of the 'Cotton Scandal' in the USSR." *Corruption and Reform* 5: 87–108.

Gleason, Gregory. 1991. "The Struggle for Control over Water in Central Asia: Republican Sovereignty and Collective Action." *Report on the USSR* 3 (June 21): 11–19.

Gleason, Gregory. 1997. *The Central Asian States: Discovering Independence.* Westview.

Gleick, Peter. 1993a. "Water and Conflict: Fresh Water Resources and International Security." *International Security* 18: 79–112.

Gleick, Peter, ed. 1993b. *Water in Crisis: A Guide to the World's Fresh Water Resources.* Oxford University Press.

Gleick, Peter. 1996. "Basic Water Requirements for Human Activities: Meeting Basic Needs." *Water International* 21: 83–92.

Gleick, Peter. 1998. *The World's Water*. Island.

Goldman, Marshall. 1992. "Environmentalism and Nationalism: An Unlikely Twist in an Unlikely Direction." In *The Soviet Environment: Problems, Policies and Politics,* ed. J. Massey Stewart. Cambridge University Press.

Goldstone, Jack. 1996. "Saving the Environment (and Political Stability Too): Institutional Responses for Developing Nations." *Environmental Change and Security Project Report,* spring: 66–71.

Gore, Albert. 1992. *Earth in the Balance*. Houghton Mifflin.

Gourevitch, Peter. 1978. "The Second Image Reversed: The International Sources of Domestic Politics." *International Organization* 32: 881–912.

Greenberg, Melanie, John Barton, and Margaret McGuinness, eds. 2000. *Words over War: Mediation and Arbitration to Prevent Deadly Conflict.* Rowman and Littefield.

Grieco, Joseph. 1993. "The Relative Gains Problem for International Cooperation." *American Political Science Review* 87 (September): 729–735.

Gustafson, Thane. 1981. *Reform in Soviet Politics: Lessons of Recent Policies on Land and Water.* Cambridge University Press.

Gustafson, Thane. 1989. *Crisis Amid Plenty: The Politics of Soviet Energy Under Brezhnev and Gorbachev.* Princeton University Press.

Gutner, Tamar. 1999. "Cleaning up the Baltic Sea: The Role of Multilateral Development Banks." In *Protecting Regional Seas: Developing Capacity and Fostering Environmental Cooperation in Europe,* ed. S. VanDeveer and G. Dabelko. Environmental Change and Security Project, Woodrow Wilson International Center for Scholars.

Haas, Ernst. 1964. *Beyond the Nation-State: Functionalism and International Organization.* Stanford University Press.

Haas, Peter M. 1992. "Banning Chlorofluorocarbons: Epistemic Community Efforts to Protect Stratospheric Ozone." *International Organization* 46, winter: 187–224.

Haas, Peter M., Robert O. Keohane, and Marc Levy, eds. 1993. *Institutions for the Earth: Sources of Effective International Environmental Protection.* MIT Press.

Haggard, Stephen, and Beth Simmons. 1987. "Theories of International Regimes." *International Organization* 41: 491–517.

Hardin, Garrett. 1968. "The Tragedy of the Commons." *Science* 162: 1243–1248.

Hardin, Russell. 1982. *Collective Action.* Johns Hopkins University Press.

Harza Engineering Company. 1993. "Evaluation of the Hydroelectric Development Program of Kyrgyzstan." Prepared for United States Agency for International Development. November.

Hauge, Wenche, and Tanja Ellingsen. 1998. "Beyond Environmental Scarcity: Causal Pathways to Conflict." *Journal of Peace Research* 35: 299–317.

Hauslohner, Peter. 1987. "Gorbachev's Social Contract." *Soviet Economy* 3: 54–89.

Helman, Gerald, and Steven Ratner. 1992–93. "Saving Failed States." *Foreign Policy* 89, winter: 3–20.

Hirsch, Francine. 2000. "Toward an Empire of Nations: Border-Making and the Formation of Soviet National Identities." *Russian Review* 59: 201–226.

Homer-Dixon, Thomas. 1994. "Environmental Scarcities and Violent Conflict: Evidence from Cases." *International Security* 19: 5–40.

Homer-Dixon, Thomas, and Jessica Blitt, eds. 1998. *Ecoviolence: Links among Environment, Population and Security.* Rowman & Littlefield.

Homer-Dixon, Thomas, and Valerie Percival. 1996. *Environmental Scarcity and Violent Conflict: Briefing Book.* American Association for the Advancement of Science and University of Toronto.

Homer-Dixon, Thomas, Jeffrey Boutwell, and George Rathjens. 1993. "Environmental Change and Violent Conflict." *Scientific American,* February: 38–45.

Huntington, Samuel P. 1973. "Transnational Organizations in World Politics." *World Politics* 25: 333–368.

Hurrell, Andrew. 1994. "A Crisis of Ecological Viability? Global Environmental Change and the Nation State." *Political Studies* 42: 146–165.

Hurrell, Andrew. 1995. "Explaining the Resurgence of Regionalism in World Politics." *Review of International Studies* 21: 331–358.

Hurrell, Andrew, and Benedict Kingsbury, eds. 1992. *The International Politics of the Environment.* Clarendon.

IBRD (International Bank for Reconstruction and Development). 1997. *World Development Report 1997: The State in a Changing World.* Oxford University Press.

ICAS (Interstate Council for the Aral Sea). 1996a. *Basic Provisions for the Development of the National Water Management Strategy of the Republic of Kazakhstan.* Aral Sea Program—Project 1.1. Report by the National Group of Kazakhstan. Draft. Almaty, Kazakhstan, January.

ICAS. 1996b. *Basic Provisions for the Development of the National Water Management Strategy of the Republic of Kyrgyz Republic.* Aral Sea Program—Project 1.1. Report by the National Group of Kyrgyz Republic. Draft. Bishkek, Kyrgyz Republic, January.

ICAS. 1996c. *Basic Provisions for the Development of the National Water Management Strategy of the Republic of Turkmenistan.* Aral Sea Program—Project 1.1. Report by the National Group of Kazakhstan. Draft. Ashgabat, Turkmenistan, January.

ICAS. 1996d. Basic Provisions for the Development of the National Water Management Strategy of the Republic of Uzbekistan. Aral Sea Program—Project 1.1. Report by the National Group of Uzbekistan. Draft. Toshkent, Uzbekistan, January.

ICAS. 1996d. Fundamental Provisions of Water Management Strategy in the Aral Sea Basin: Common Strategy of Water Allocation, Rational Water Use and Protection of Water Resources. Developed with the Assistance of the World Bank for Kazakhstan, Kyrgyz Republic, Tadjikistan, Turkmenistan, and Uzbekistan.

Igamberdiyev, R., and T. Abdurakhmanova. 1975. *Istoriia rasvitiia irrigatsii v Uzbekistan. 1925–1937.* FAN.

Ilkhamov, Alisher. 1998. "*Shirkats, Dekhqon,* Farmers and Others: Farm Restructuring in Uzbekistan." *Central Asian Survey* 17: 539–560.

IMF (International Monetary Fund). 1992. *Economic Review: Uzbekistan.*

IMF. 1994. *Turkmenistan.*

ISAR (Initiative for Social Action and Renewal in Eurasia). 1999. "Summary Report to the Turner Foundation: Supporting Alternative Energy in Central Asia." February.

Iskandarov, Khasan. 1994. "Will Toshkent become a Port City?" *Current Digest of the Soviet Press* 46: 26.

Jackson, Robert. 1990. *Quasi-States: Sovereignty, International Relations and the Third World.* Cambridge University Press.

Jackson, Robert. 1993. "The Weight of Ideas in Decolonization: Normative Change in International Relations." In *Ideas and Foreign Policy: Beliefs, Institutions, and Political Change,* ed. J. Goldstein and R. Keohane. Cornell University Press.

Jackson, Robert, and Carl Rosberg. 1982. "Why Africa's Weak States Persist: The Empirical and Juridical in Statehood." *World Politics* 35 (October): 1–24.

Jacobson, Harold. 1984. *Networks of Interdependence: International Organizations and the Global Political System.* Knopf.

Joffe, Muriel. 1984. "Regional Rivalry and Economic Nationalism: The Central Industrial Regional Industrialists' Strategy for the Development of the Russian Economy, 1880s–1914." *Russian History* 11: 389–421.

Joffe, Muriel. 1995. "Autocracy, Capitalism and Empire: The Politics of Irrigation." *Russian Review* 54 (July): 365–388.

Jones, Rodney, and Mark McDonough. 1998. *Tracking Nuclear Proliferation: A Guide in Maps and Charts.* Carnegie Endowment for International Peace.

Jones Luong, Pauline. 1997. Ethno-politics and Institutional Design: Explaining the Establishment of Electoral Systems in Post-Soviet Central Asia. Ph.D. Dissertation. Harvard University.

Jones Luong, Pauline. 2000. "After the Break-up: Institutional Design in Transitional States." *Comparative Political Studies* 33: 563–592.

Jones Luong, Pauline, and Erika Weinthal. 1999. "The NGO Paradox: Goals, Strategies, and Non-Democratic Outcomes in Kazakhstan." *Europe-Asia Studies* 51: 1267–1284.

Jones Luong, Pauline, and Erika Weinthal. 2001. "Prelude to the Resource Curse: Oil and Gas Development Strategies in Central Asia and Beyond." *Comparative Political Studies,* May: 367–399.

Kaiser, Robert. *The Geography of Nationalism in Russia and the USSR.* Princeton University Press. 1994.

Kamalov, Yusup. 1996. "Civilization by the Aral Sea Verges on Extinction." *Surviving Together,* summer: 24–26.

Kamieniecki, Sheldon, ed. 1993. *Environmental Politics in the International Arena: Movements, Parties, Organizations, and Policy.* State University of New York Press.

Kapuscinski, Ryszard. 1994. *Imperium.* Vintage Books.

Karimov, Akhmal. 1998. "Water Regimes in Central Asia." Prepared for Aral Sea Basin Workshop sponsored by the Social Science Research Council, May 19–21, Toshkent, Uzbekistan.

Karl, Terry Lynn. 1997. *The Paradox of Plenty: Oil Booms and Petro-States.* University of California Press.

Keck, Margaret, and Kathryn Sikkink. 1998. *Activists beyond Borders: Advocacy Networks in International Politics*. Cornell University Press.

Kennedy, Donald, David Holloway, Erika Weinthal, Walton Falcon, Paul Ehrlich, Rosamond Naylor, Michael May, Steven Schneider, Steven Fetter, and Jor-san Choi. 1998. "Environmental Quality and Regional Security." Carnegie Commission on Preventing Deadly Conflict, December.

Keohane, Robert. 1982. "The Demand for International Regimes." *International Organization* 36: 325–355.

Keohane, Robert. 1984. *After Hegemony: Cooperation and Discord in the World Political Economy*. Princeton University Press.

Keohane, Robert, ed. 1986a. *Neorealism and Its Critics*. Columbia University Press.

Keohane, Robert. 1986b. "Reciprocity in International Relations." *International Organization* 40 (winter): 1–27.

Keohane, Robert. 1993. "Institutional Theory and the Realist Challenge After the Cold War." In *Neorealism and Neoliberalism: The Contemporary Debate*, ed. D. Baldwin. Columbia University Press.

Keohane, Robert, and Marc Levy, eds. 1996. *Institutions for Environmental Aid: Pitfalls and Promise*. MIT Press.

Keohane, Robert, and Joseph Nye Jr., eds. 1972. *Transnational Relations and World Politics*. Harvard University Press.

Keohane, Robert, and Elinor Ostrom. 1994. "Introduction to the Special Issue on Local Commons and Global Interdependence." *Journal of Theoretical Politics* 6: 403–428.

Keohane, Robert, and Elinor Ostrom, eds. 1995. *Local Commons and Global Interdependence*. Sage.

Khazanov, A. 1992. "Nomads and Oases in Central Asia." In *Transition to Modernity: Essays on Power, Wealth and Belief*, ed. J. Hall and I. Jarvie. Cambridge University Press.

Kirmani, Syed, and Guy Le Moigne. 1997. Fostering Riparian Cooperation in International River Basins. Technical Paper 335, World Bank.

Klötzli, Stefan. 1993. Der slowakisch-ungarische Konflikt um das Staustufenprojekt Gabcikovo. ENCOP Occasional Paper 7, Center for Security Policy and Conflict Research, Zurich.

Klötzli, Stefan. 1994. The Water and Soil Crisis in Central Asia—A Source for Future Conflicts? ENCOP Occasional Paper 11, Center for Security Policy and Conflict Research, Zurich.

Kolbasev, O. 1971. "Legislation on Water Use in the USSR." In *Water Resources Law and Policy in the Soviet Union*, ed. I. Fox. University of Wisconsin Press.

Kornai, János. 1992. *The Socialist System: The Political Economy of Communism*. Princeton University Press.

Krasner, Stephen. 1976. "State Power and the Structure of International Trade." *World Politics* 28: 317–345.

Kratochwil, Friedrich. 1986. "Of Systems, Boundaries, and Territoriality: An Inquiry into the Formation of the State System." *World Politics* 39: 27–52.

Laschenov, V. 1990. "Problemi mezhrespublikanskovo raspredeleniia vodnikh resursov R. Syrdari." *Melioratsiia i vodnoe khoziaistvo* 1: 3–5.

LeMarquand, David. 1977. *International Rivers: The Politics of Cooperation.* Westwater Research Centre.

Lenin and Stalin. 1940. *Stat'i i rechi o Srednei Azii i Uzbekistane: Sbornik.* Toshkent, Uzbekistan: Partizdat: Central Committee of the Communist Party of Uzbekistan.

Levintanus, Arkady. 1992. "Saving the Aral Sea." *The Environmentalist* 12: 85–91.

Levy, Marc. 1993. "European Acid Rain: The Power of Tote-Board Diplomacy." In *Institutions for the Earth: Sources of Effective International Protection,* ed. P. Haas et al. MIT Press.

Libecap, Gary. 1994. "The Conditions for Successful Collective Action." *Journal of Theoretical Politics* 6: 563–592.

Linz, Juan, and Alfred Stepan. 1996. *Problems of Democratic Transition and Consolidation: Southern Europe, South America, and Post communist Europe.* Johns Hopkins University Press.

Lipovsky, Igor. 1995. "The Central Asian Cotton Epic." *Central Asian Survey* 14: 529–542.

Lipschutz, Ronnie. 1998. "Damming Troubled Waters: Conflict over the Danube. 1950–2000." Presented at conference on Environment and Violent Conflict, Institute of War and Peace Studies, Columbia University.

Litfin, Karen. 1993. "Eco-Regimes: Playing Tug of War with the Nation-State." In *The State and Social Power in Global Environmental Politics,* ed. R. Lipschutz and K. Conca. Columbia University Press.

Lowi, Miriam. 1993a. *Water and Power: The Politics of a Scarce Resource in the Jordan River Basin.* Cambridge University Press.

Lowi, Miriam. 1993b. "Bridging the Divide: Transboundary Resource Disputes and the Case of West Bank Water." *International Security* 18: 113–138.

Lubin, Nancy. 1984. *Labour and Nationality in Soviet Central Asia.* Princeton University Press.

Lyons, Gene, and Michael Mastanduno, eds. 1995. *Beyond Westphalia? State Sovereignty and International Intervention.* Johns Hopkins University Press.

Mansfield, Edward, and Jack Snyder. 1995. "Democratization and the Danger of War." *International Security* 20, summer: 5–38.

Maoz, Zeev. 1996. *Domestic Sources of Global Change.* University of Michigan Press.

Martin, Lisa. 1994. "Heterogeneity, Linkage and Commons Problems." *Journal of Theoretical Politics* 6: 473–493.

Masanov, Nurbulat. 1995. *Kochevaia sivilizatsiia Kazakhov*. Almaty, Kazakhstan: Sotsinvest.

Mathews, Jessica Tuchman. 1989. "Redefining Security." *Foreign Affairs* 68, spring: 161–177.

Mathews, Jessica. 1997. "Power Shift." *Foreign Affairs* 76, January-February: 50–66.

Matley, Ian. 1994. "Agricultural Development (1865–1963)." In *Central Asia: 130 Years of Russian Dominance,* ed. E. Allworth. Duke University Press.

Matthew, Richard. 1999. "Scarcity and Security: A Common-Pool Resource Perspective." In *Anarchy and the Environment,* ed. J. Barkin and G. Shambaugh. State University of New York Press.

Matthews, John. 1996. "Current Gains and Future Outcomes: When Cumulative Relative Gains Matter." *International Security* 21: 112–146.

Mayer, Frederick. 1992. "Managing Domestic Differences in International Negotiations: The Strategic Use of Internal Side-payments." *International Organization.* 46: 793–818.

McCaffrey, Stephen. 1993. "Water, Politics, and International Law." In *Water in Crisis: A Guide to the World's Fresh Water Resources,* ed. P. Gleick. Oxford University Press.

McKinney, Daene, and Sandra Akmansoy. 1998. "What are the Competing Water Needs and Uses in the Aral Region?" Presented at SSRC Conference in Toshkent, Uzbekistan, May 19–21.

McNeely, Connie L. 1995. *Constructing the Nation-State: International Organization and Prescriptive Action.* Greenwood.

Mendelson, Sarah, and John Glenn. 2000. Democracy Assistance and NGO Strategies in Post-Communist Societies. Working Paper 8, Carnegie Endowment.

Meyer, John, and Brian Rowan. 1991. "Institutionalized Organizations: Formal Structure as Myth and Ceremony." In *The New Institutionalism in Organizational Analysis,* ed. W. Powell and P. DiMaggio. University of Chicago Press.

Michel, Aloys A. 1967. *The Indus Rivers: A Study of the Effects of Partition.* Yale University Press.

Micklin, Philip P. 1991. The Water Management Crisis in Soviet Central Asia. No. 905, Carl Beck Papers, Center for Russian and East European Studies. University of Pittsburgh.

Micklin, Philip P. 1992a. "The Aral Crisis: Introduction to the Special Issue." *Post-Soviet Geography* 33: 269–282.

Micklin, Philip P. 1992b. "Water Management in Soviet Central Asia: Problems and Prospects." In *The Soviet Environment: Problems, Policies and Politics,* ed. J. Stewart. Cambridge University Press.

Micklin, Philip P. 1997a. Final Project Report to the Government of Uzbekistan. Prepared for the Central Asian Mission, USAID. Contract No. CCN-0003-Q-14–3165-00.

Micklin, Philip P. 1997b. "Draft Final Report on Training Seminar: Developing Water Pricing Systems for Uzbekistan." Prepared for the Central Asian Mission, USAID. Contract No. CCN-0003-Q-14–3165–00, 11 August.

Micklin, Philip P. 1998. "International and Regional Responses to the Aral Sea Crisis: An Overview of Efforts and Accomplishments." Presented at Aral Sea Basin Water Management Workshop sponsored by Social Science Research Council, Toshkent, Uzbekistan, May 19–21.

Micklin, Philip P., and Andrew Bond. 1988. "Reflections on Environmentalism and the River Diversion Projects." *Soviet Economy* 4: 253–274.

Migdal, Joel. 1988. *Strong Societies, Weak States: State-Society Relations and State Capabilities in the Third World*. Princeton University Press.

Milgrom, Paul, and John Roberts. 1992. *Economics, Organizations, and Management*. Prentice-Hall.

Milner, Helen. 1992. "International Theories of Cooperation Among Nations: Strengths and Weaknesses." *World Politics* 44: 466–496.

Mitchell, Ronald. 1994. *Intentional Oil Pollution at Sea: Environmental Policy and Treaty Compliance*. MIT Press.

Mitrany, David. 1966. *A Working Peace System*. Quandrangle.

Moore, Jonathan, ed. 1998. *Hard Choices: Moral Dilemmas in Humanitarian Intervention*. Rowman & Littlefield.

Moravcsik, Andrew. 1993. "Introduction: Integrating International and Domestic Theories of International Bargaining." In *Double-Edged Diplomacy: International Bargaining and Domestic Politics*, ed. P. Evans et al. University of California Press.

Moravcsik, Andrew. 1994. "Why the European Community Strengthens the State: Domestic Politics and International Cooperation." Presented at Annual Meeting of American Political Science Association, New York.

Munro-Faure, Paul. 1994. Interim Report on Land Reform Issues, May-June. ULG Consultants Ltd, The Ministry of Agriculture of the Government of Kyrgyzstan and EC-TACIS Sectoral Support Program.

Myers, Norman. 1993. *Ultimate Security*. Norton.

Nanni, Marcella. 1996. "The Aral Sea Basin: Legal and Institutional Issues." *Review of European Community and International Environmental Law* 5: 130–137.

Naumkin, Vitaly, ed. 1994. *Central Asia and Transcaucasia: Ethnicity and Conflict*. Greenwood.

Nelson, Joan. 1993. "The Politics of Economic Transformation: Is the Third World Experience Relevant in Eastern Europe?" *World Politics* 45 (April): 433–463.

North, Douglass. 1990. *Institutions, Institutional Change and Economic Performance.* Cambridge University Press.

Nunn, Sen. Sam, Nancy Lubin, and Barnett Rubin. 1999. *Calming the Ferghana Valley: Development and Dialogue in the Heart of Central Asia.* Council on Foreign Relations and Century Foundation.

O'Donnell, Guillermo, and Philippe Schmitter. 1986. *Transitions from Authoritarian Rule: Tentative Conclusions about Uncertain Democracies.* Johns Hopkins University Press.

Ohlsson, Leif, ed. 1995. *Hydropolitics: Conflicts over Water as a Development Constraint.* Zed Books.

Olcott, Martha Brill. 1993. "Central Asia on Its Own." *Journal of Democracy* 4 (January): 92–103.

Olcott, Martha Brill. 1996. *Central Asia's New States: Independence, Foreign Policy, and Regional Security.* United Institute of Peace Press.

Ostrom, Elinor. 1990. *Governing the Commons: The Evolution of Institutions for Collective Action.* Cambridge University Press.

Ostrom, Elinor. 1992. *Crafting Institutions for Self-Governing Irrigation Systems.* ICS.

Ostrom, Elinor. 1994. "Constituting Social Capital and Collective Action." *Journal of Theoretical Politics* 6: 527–562.

Ostrom, Elinor, Roy Gardner, and James Walker. 1994. *Rules, Games, and Common-Pool Resources.* University of Michigan Press.

Oye, Kenneth, ed. 1986. *Cooperation under Anarchy.* Princeton University Press.

Palen, K. 1910. *Oroshenie v Turkestane.* Senate Publishing House.

Panarin, Sergei 1994. "Political Dynamics of the 'New East' (1985–1993)." In *Central Asia and Transcaucasia: Ethnicity and Conflict,* ed. V. Naumkin. Greenwood.

Perry, William. 1996. "Defense in an Age of Hope." *Foreign Affairs* 75: 64–79.

Peterson, D. 1993. *Troubled Lands: The Legacy of Soviet Environmental Destruction.* Westview.

Pierce, Richard. 1964. *Mission to Turkestan: Being the Memoirs of Count K. K. Pahlen, 1908–1909.* Oxford University Press.

Polat, Abdumannob. 1995. "Central Asian Security Forces Against Their Dissidents in Exile." In *Central Asia: Conflict, Resolution, and Change,* ed. R. Sagdeev and S. Eisenhower. CPSS.

Pomfret, Richard. 1995. *The Economies of Central Asia.* Princeton University Press.

Porter, Gareth, and Janet Brown. 1991. *Global Environmental Politics.* Westview.

Postel, Sandra. 1993. "Water and Agriculture." In *Water in Crisis,* ed. P. Gleick. Oxford University Press.

Postel, Sandra. 1996. Dividing the Waters: Food Security, Ecosystem Health, and the New Politics of Scarcity. Paper 132, Worldwatch Institute.

Princen, Thomas, and Matthias Finger. 1994. *Environmental NGOs in World Politics: Linking the Local and the Global.* Routledge.

Prosser, Sarah. 2000. "Reform within and without the Law: Further Challenges for Central Asian NGOs." *Harvard Asia Quarterly* 4, no. 3.

Przeworski, Adam. 1991. *Democracy and the Market.* Cambridge University Press.

Putnam, Robert. 1988. "Diplomacy and Domestic Politics: The Logic of the Two-Level Game." *International Organization* 42: 427–460.

Raiffa, Howard. 1982. *The Art and Science of Negotiation.* Harvard University Press.

Rakhimov, E. 1990. *Sotsialno-ekonomicheskie problemy Arala i Priaralia.* FAN.

Rassudova, R. 1969. "Zaniatiia naseleniia." In *Etnograficheskie ocherki Uzbekskogo selskogo naseleniia,* ed. G. Vasileva and B. Karmisheva. Nauka.

Reinicke, Wolfgang. 1996. "Can International Financial Institutions Prevent Internal Violence? The Sources of Ethno-National Conflict in Transitional Societies." In *Preventing Conflict in the Post-Communist World,* ed. A. Chayes and A. Handler Chayes. Brookings Institution.

Reznichenko, Grigori. 1992. *The Aral Sea Tragedy.* Novosti.

Risse-Kappen, Thomas, ed. 1995. *Bringing Transnational Relations Back In: Non-state Actors, Domestic Structures, and International Institutions.* Cambridge University Press.

Rochester, J. 1986. "The Rise and Fall of International Organization as a Field of Study." *International Organization* 40: 777–813.

Rosenau, James N. 1986. "Before Cooperation: Hegemons, Regimes, and Habit-Driven Actors in World Politics." *International Organization* 40: 849–894.

Rosenthal, Jean-Laurent. 1992. *The Fruits of the Revolution: Property Rights, Litigation, and French Agriculture, 1700–1860.* Cambridge University Press.

Rostankowski, Peter. 1982. "Transformation of Nature in the Soviet Union: Proposals, Plans and Reality." *Soviet Geography* 22 (June): 381–390.

Rothchild, Donald, and Naomi Chazan, eds. 1988. *The Precarious Balance: State and Society in Africa.* Westview.

Rubin, Barnett. 1995. *The Fragmentation of Afghanistan.* Yale University Press.

Ruggie, John Gerard. 1992. "Multilaterialism: The Anatomy of an Institution." *International Organization* 46, summer: 561–598.

Rumer, Boris. 1989. *Soviet Central Asia: A Tragic Experiment.* Unwin Hyman.

Schelling, Thomas. 1960. *The Strategy of Conflict.* Harvard University Press.

Schmitter, Philippe, and Terry Lynn Karl. 1994. "The Conceptual Travels of Transitologists and Consolidologists: How Far to the East Should They Attempt to Go?" *Slavic Review* 53, spring: 173–185.

Schreurs, Miranda, and Elizabeth Economy, eds. 1997. *The Internationalization of Environmental Protection.* Cambridge University Press.

Scott, James. 1998. *Seeing Like a State: How Certain Schemes to Improve the Human Condition Have Failed.* Yale University Press.

Sebenius, James. 1983. "Negotiation Arithmetic: Adding and Subtracting Issues and Parties." *International Organization* 37, spring: 281–316.

Severin, Barbara. 1987. "Special Report on Soviet Cotton Production Data." *Research on Soviet and East European Agriculture* 9, December: 1–2.

Shiklomanov, Igor. 1993. "World Fresh Water Resources." In *Water in Crisis,* ed. P. Gleick. Oxford University Press.

Shim, Ui-Sup. 1995. Transition to Market Economy in the Central Asian Republics: Korean Community and Market Economy. Institute of Developing Economies, Japan.

Sinnott, Pete. 1992r. "The Physical Geography of Soviet Central Asia and the Aral Sea Problem." In *Geographic Perspectives on Soviet Central Asia,* ed. R. Lewis. Routledge.

Sirozhidnikov, K. 1991. "Ob vyiavlennykh prichinakh snizheniia urovnia Aralskogo Moria." *Problemy osvoeyeniia pustyn* 6: 23–27.

Skocpol, Theda. 1979. *States and Social Revolutions.* Cambridge University Press.

Skocpol, Theda. 1985. "Bringing the State Back In: Strategies of Analysis in Current Research. In *Bringing the State Back In,* ed. P. Evans et al. Cambridge University Press.

Slezkine, Yuri. 1994. "The USSR as a Communal Apartment, or How a Socialist State Promoted Ethnic Particularism." *Slavic Review* 53, summer: 414–452.

Smith, David. 1995. "Environmental Security and Shared Water Resources in Post-Soviet Central Asia." *Post-Soviet Geography* 36: 351–370.

Smith, David. 1994. "Change and Variability in Climate and Ecosystem Decline in Aral Sea Basin Deltas." *Post-Soviet Geography* 35: 142–265.

Solodennikof, D. 1996. "Issues of the Management of the Toktogul Reservoir Under Current Conditions of a Joint Use of the Syr-Darya Basin Water Resources by Kyrgyzstan, Uzbekistan, Tajikistan, and Kazakhstan." *Aral Herald (Central Asian Scientific Tribune)* 1, spring: 17–22.

Spoor, Max. 1995. "Agrarian Transition in Former Soviet Central Asia: A Comparative Study of Uzbekistan and Kyrgyzstan." *Journal of Peasant Studies* 23, October: 46–63.

Stark, David. 1992. "Path Dependence and Privatization Strategies in East Central Europe." *East European Politics and Societies* 4: 17–53.

Starr, S. 1996. "Making Eurasia Stable." *Foreign Affairs* 75, January-February: 80–92.

Stein, Melanie. 1996. "Conflict Prevention in Transition Economies: A Role for the European Bank for Reconstruction and Development." In *Preventing Conflict in the Post-Communist World*, ed. A. Chayes and A. Handler Chayes. Brookings Institution.

Suny, Ronald Grigor. 1995. "Ambiguous Categories: States, Empires and Nations." *Post-Soviet Affairs* 11, April–June: 185–196.

Susskind, Lawrence. 1994. *Environmental Diplomacy: Negotiating More Effective Global Agreements*. Oxford University Press.

Taylor, Michael. 1987. *The Possibility of Cooperation*. Cambridge University Press.

Taylor, Michael, and Sara Singleton. 1993. "The Communal Resource: Transaction Costs and the Solution of Collective Action Problems." *Politics and Society* 21: 195–214.

Teclaff, Ludwik. 1967. *The River Basin in History and Law*. Martinus Nijhoff.

Tetlock, Philip, and Aaron Belkin, eds. 1996. *Counterfactual Thought Experiments in World Politics: Logical, Methodological, and Psychological Perspectives*. Princeton University Press.

Thelen, Kathleen, and Sven Steinmo. 1992. "Historical Institutionalism in Comparative Politics." In *Structuring Politics*, ed. S. Steinmo et al. Cambridge University Press.

Thomson, Janice. 1995. "State Sovereignty in International Relations: Bridging the Gap Between Theory and Empirical Research." *International Studies Quarterly* 39: 213–233.

Thurman, Michael. 1995. "Irrigation Management in Uzbekistan: Economic Inefficiencies, Costs, and Possible Solutions." *Central Asian Monitor* 4: 24–37.

Tilly, Charles. 1975. *The Formation of National States in Europe*. Princeton University Press.

Tilly, Charles. 1990. *Coercion, Capital and European States, AD 990–1990*. Blackwell.

Tishkov, Valery. 1995. "Don't Kill Me, I'm a Kyrgyz!" *Journal of Peace Research* 32, May 2: 133–149.

Tolba, Mostafa, with Iwona Rummel-Bulska. 1998. *Global Environmental Diplomacy: Negotiating Environmental Agreements for the World. 1973–1992*. MIT Press.

Tsaruk, Oleg. 1998. Formal Response to Philip Micklin's Paper at SSRC Water Conference. May 19–21, Toshkent, Uzbekistan.

Ullman, Richard. 1983. "Redefining Security." *International Security* 8, summer: 129–153.

United Nations. 1978. *Register of International Rivers*. Pergamon.

VanDeveer, Stacy. 2000. "Protecting Europe's Seas: Lessons from the Last 25 Years." *Environment* 42, July-August: 10–26.

Vasileva, G., and B. Karmisheva, eds. 1969. *Etnograficheskie ocherki Uzbekskogo selskogo naseleniia*. Nauka.

Victor, David, Kal Raustiala, and Eugene Skolnikoff, eds. 1998. *The Implementation and Effectiveness of International Environmental Commitments: Theory and Practice.* MIT Press.

Vinogradov, Sergei. 1996. "Transboundary Water Resources in the Former Soviet Union: Between Conflict and Cooperation." *Natural Resources Journal:* 393–415.

Walker, Martin. 1986. *The Waking Giant: Gorbachev's Russia.* Pantheon Books.

Waltz, Kenneth N. 1979. *Theory of International Politics.* Addison-Wesley.

Wapner, Paul. 1995. "Politics beyond the State: Environmental Activism and World Civic Politics." *World Politics* 47, April: 311–340.

Waterbury, John. 1994. "Transboundary Water and the Challenge of International Cooperation in the Middle East." In *Water in the Arab World,* ed. P. Rogers and P. Lydon. Harvard University Press.

Weber, Max. 1964. *The Theory of Social and Economic Organization,* ed. T. Parsons. Free Press.

Wedel, Janine. 1998. *Collision and Collusion: The Strange Case of Western Aid to Eastern Europe 1989–1998.* St. Martin's Press.

Weiner, Douglas. 1988. *Models of Nature: Conservation, Ecology, and Cultural Revolution.* Indiana University Press.

Weinthal, Erika. 2000. "Making Waves: Third Parties and International Mediation in the Aral Sea Basin." In *Words over War,* ed. M. Greenberg et al. Rowman and Littefield.

Weinthal, Erika. 2001. "Sins of Omission: Constructing Negotiating Sets in the Aral Sea Basin." *Journal of Environment and Development* 10, March: 50–79.

Weinthal, Erika, and Pauline Jones Luong. 2000. "Weak State in Formation? Energy Wealth and Tax Reform in Kazakhstan." Presented at Annual Meeting of American Political Science Association, Washington.

Weiss, Edith Brown, and Harold Jacobson, eds. 1998. *Engaging Countries: Strengthening Compliance with International Environmental Accords.* MIT Press.

Wheeler, Geoffrey. 1966. *The Peoples of Soviet Central Asia.* Bodley Head.

Winner, Irene. 1963a. "Some Problems on Nomadism and Social Organization among the Recently Settled Kazakhs: Part 1." *Central Asian Review* 11: 246–267.

Winner, Irene. 1963b. "Some Problems on Nomadism and Social Organization among the Recently Settled Kazakhs: Part 2." *Central Asian Review* 11: 355–373.

Wolfson, Ze'ev. 1990. "Central Asian Environment: A Dead End." *Environmental Policy Review* 4: 29–46.

Wood, Major Herbert. 1876. *The Shores of Lake Aral.* Smith, Elder.

World Bank. 1993a. *The Aral Sea Crisis: Proposed Framework of Activities.*

World Bank. 1993b. *Uzbekistan: An Agenda for Economic Reform.*

World Bank. 1994a. *Aral Sea Program—Phase 1, Aide Memoire, Volume 2,* Work Bank Preparation Mission.

World Bank. 1994b. *Turkmenistan.*

World Bank. 1995a. *Aral Sea Basin Program—Phase 1, Progress Report 1.*

World Bank. 1995b. *Aral Sea Basin Program—Phase 1, Progress Report 2.*

World Bank. 1996a. "Aide Memoire—Visit of Mr. Peter Whitford."

World Bank. 1996b. *Aral Sea Basin Program—Phase 1, Progress Report 3.* February.

World Bank. 1997. "Aral Sea Basin Program Review: Proposed Tentative Conclusions and Recommendations." March.

World Bank. 1998. Aral Sea Basin Program—Water and Environmental Management Project, Project Document, Volume 1—Main Report. Report 17587-UZ.

World Resources Institute, United Nations Environment Programme, United Nations Development Programme, and World Bank. 1996. *World Resources: A Guide to the Global Environment, 1996–97.* Oxford University Press.

Young, Oran. 1994a. *International Governance: Protection the Environment in a Stateless Society.* Cornell University Press.

Young, Oran. 1994b. "The Problem of Scale in Human/Environment Relationships." *Journal of Theoretical Politics* 6: 429–447.

Ziegler, Charles. 1987. *Environmental Policy in the USSR.* University of Massachusetts Press.

Zile, Zigurds L. 1971. "Kolbasov's Legislation on Water Use in the USSR from the Perspective of Recent Trends in Soviet Law." In *Water Resources Law and Policy in the Soviet Union,* ed. I. Fox. University of Wisconsin Press.

Zürn, Michael. "The Rise of International Environmental Politics." *World Politics* 50 (1998): 617–649.

Index